Advances in Anatomy
Embryology and Cell Biology

Vol. 129

Editors

F. Beck, Melbourne W. Hild, Galveston
W. Kriz, Heidelberg J.E. Pauly, Little Rock
Y. Sano, Kyoto T.H. Schiebler, Würzburg

Martin Oudega   Egbert A.J.F. Lakke
Enrico Marani   Raph. T.W.M. Thomeer

Development of the Rat Spinal Cord: Immuno- and Enzyme Histochemical Approaches

With 33 Figures

Springer-Verlag
Berlin Heidelberg New York
London Paris Tokyo
Hong Kong Barcelona
Budapest

M. Oudega, Ph.D
The Miami Project to Cure Paralysis
University of Miami, School of Medicine
1600 NW 10th Avenue, R-48
Miami, FL 33136, USA

E. Lakke, M.D.
E. Marani, Ph.D
Neuroregulation Group, Department of Physiology
Leiden University
P.O. Box 9604
2300 RC Leiden
The Netherlands

R. Thomeer, M.D., Ph.D
Professor of Neurosurgery
Neuroregulation Group, Department of Neurosurgery
Leiden Academic Hospital
P.O. Box 9600
2300 RC Leiden
The Netherlands

ISBN 3-540-57173-6 Springer-Verlag Berlin Heidelberg New York
ISBN 0-387-57173-6 Springer-Verlag New York Berlin Heidelberg

CIP data applied for

Typesetting: Best-set Typesetter Ltd., Hong Kong
27/3130/SPS-5 4 3 2 1 0 – Printed on acid-free paper

Preface

The studies described here were carried out in the Neuroregulation Group, Department of Physiology, University of Leiden, the Netherlands. Over the last decade, this group, in close collaboration with the Department of Neurosurgery of the Academic Hospital of Leiden, has studied the development of the central nervous system from a neuroanatomical as well as a clinical perspective.

During this period, the expression of several morphoregulators in the developing rat spinal cord was extensively investigated. Parallel studies focused on the development of the spinal cord fiber systems, which was studied by means of the intrauterine use of neuronal tracers. The main goal of these studies was to extend our knowledge about the (normal) generation of the spinal cord and to contribute to the understanding of clinical problems related to regeneration and degeneration in the mammalian central nervous system. The studies on morphoregulators, in particular, appeared to benefit two different scientific areas. Firstly, the correlation between morphoregulator expression patterns and known anatomy contributed to our knowledge about spinal cord development. Secondly, the correlation between morphoregulator expression patterns and known developmental processes may help to understand their precise function(s). This volume of *Advances in Anatomy, Embryology and Cell Biology* presents these particular studies on the development of the rat spinal cord performed over the last decade. As well as integrating the results of the tracer studies, this volume also provides an update on the development of the rat spinal cord.

This study would not have been possible without the help and support of several people. We are grateful to Dr. W.J.T. Wessels for her invaluable contribution to this study and would like to thank M.G.M. Deenen, J.G.P.M. van der Veeken, J.M. Guldemond, and Dr. J. Dijk for their excellent technical assistance. We are also grateful to Dr. A. Matus (Zürich, Switzerland) for his generous gift of the MAP 2 antibody, and much appreciate the helpful comments of, and discussions with, Dr. J.H.C. Voormolen. We would like to express our special thanks to Dr. J.K. Mai (Düsseldorf,

Germany) and Dr. B.M. Riederer (Lausanne, Switzerland) for their critical reading, comments, and discussions on parts of the manuscript. Finally, we are indebted to Dr. J. Voogd (Rotterdam, the Netherlands) for the critical evaluation of different parts of the manuscript.

Miami, Leiden/November 1993

Martin Oudega
Egbert Lakke
Enrico Marani
Raph Thomeer

Contents

1 Introduction

The spinal cord has been the subject of intensive neuroanatomical and morphological research for more than a century (His 1889; Ramon Y Cajal 1890), mainly because it provided an appropriate model to study the relationship between (neural) anatomy and function. The anatomical organization of the rat spinal cord is relatively simple and forms the basis for its function as an intermediate between the central brain and the peripheral body.

In the spinal cord, the outgoing motor information is conveyed by the axons of the motor neurons which are arranged in the motor columns (Rexed's lamina IX; Rexed 1954) in the ventral horn (see Fig. 1). The axons of the motor neurons course along the motor roots and leave the cord at its ventrolateral aspect to innervate the skeletal muscles of the body and limbs (Kuhlenbeck 1975). Central (motor) commands reach the spinal motor neurons via the descending supraspinal pathways. Besides this controlled voluntary movement, an involuntary reflex pathway can also be generated in the spinal cord (Edinger 1912). A reflex conveys sensory input from the periphery directly or indirectly to the spinal motor columns (Eccles and Pritchard 1937; Renshaw 1940; Kuhlenbeck 1975; Sterling and Kuypers 1967).

The peripheral sensory information is conducted via the dorsal roots, the bundled, centrally projecting axons of the spinal ganglion cells (Brodal 1981). These roots enter the spinal cord at its dorsolateral aspect. Most of the incoming fibers make contact with the neurons of the upper three layers of the dorsal horn (Rexed's laminae I–III; Rexed 1954; see Fig. 1). Part of the incoming information is transmitted directly to higher brain regions (Giesler et al. 1978, 1979; Menétrey et al. 1982; Liu 1983; Morrel and Pfaff 1983). Thicker myelinated dorsal root fibers ascend in the dorsal funiculus, whereas the thinner unmyelinated primary afferents ascend in the dorsolateral fasciculus towards rostral spinal cord levels or higher brain regions (Brown 1981).

The intermediate gray (Rexed's laminae IV–VIII; Rexed 1954; see Fig. 1) contains the ipsilaterally and contralaterally projecting neurons which convey sensory and motor information either directly to the periphery or towards different brain areas, especially the brain stem, the thalamus, and the cerebellum (Verburgh and Kuypers 1987). According to their function, the neurons are often referred to as relay neurons. The intermediate gray also contains the interneurons whose axons run in the marginal layer close to the gray matter, the so-called fasciculus proprius (also known as the cornu marginal zone). They re-enter the gray matter a few levels higher or lower in the cord to contact other interneurons or motor neurons (Molenaar and Kuypers 1978). Thus,

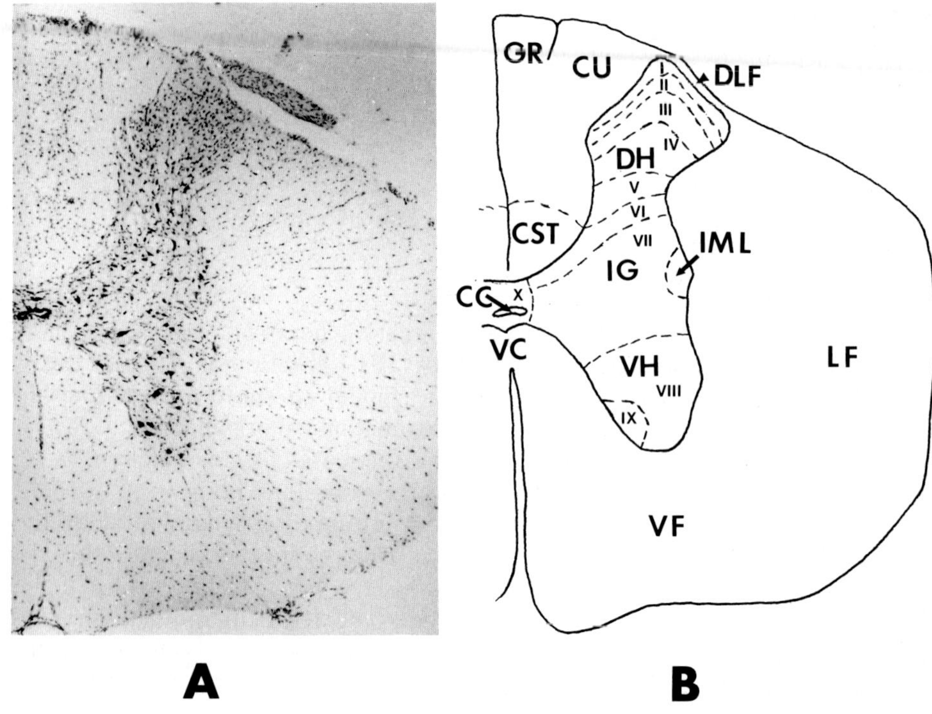

Fig. 1A,B. Transversal representations of the adult rat thoracic spinal cord. **A** Cresyl violet-stained section. **B** Schematic drawing of the same section as in **A**. Indicated are the different recognizable structures and Rexed's laminae (*I–X*; Rexed 1954). *DH*, dorsal horn; *CST*, corticospinal tract; *IG*, intermediate gray; *CC*, central canal; *VC*, ventral commissure; *VH*, ventral horn; *LF*, lateral funiculus; *VF*, ventral funiculus; *CU*, fasciculus cuneatus; *DLF*, dorsolateral fasciculus; *GR*, fasciculus gracilis; *IML*, intermediolateral cell column

these axons provide the intrinsic connections between the different spinal cord levels and are known as propriospinal fibers (Kuhlenbeck 1975; Brown 1981).

The supraspinal axons that connect the spinal cord with the higher brain center and vice versa are located in the marginal layer, which is divided into several funiculi (see Kuhlenbeck 1975). The ventral and lateral funiculi contain the ascending fiber systems of the contralaterally and ipsilaterally projecting neurons of the intermediate gray and the descending fiber tracts from several higher brain nuclei (Kuypers 1981). In addition, they enclose the propriospinal fibers of the cornu marginal zone (Kuhlenbeck 1975). The dorsolateral fasciculus of the rat spinal cord contains the incoming thinner, unmyelinated sensory fibers of the dorsal root ganglion cells. The thicker, myelinated fibers of these particular cells project their ascending branches into the dorsal funiculus. The most ventral part of the dorsal funiculus encloses the descending fibers of the corticospinal tract (Brown 1971; Kuypers 1981; Gribnau et al. 1986).

The plain anatomical organization of the spinal cord does not hinder the performance of its functions. The spinal cord functions as an interface in conducting the incoming sensory and outgoing motor information by way of the spinal nerves, which also convey the information that regulates and controls the sympathetic and parasympathetic processes of the organism. In addition, the spinal cord functions as an integration center of information from supraspinal, spinal, and peripheral levels. Thus, the spinal cord represents an active, intermediate structure between the central brain and the peripheral environment.

Because of its straightforward organization, the spinal cord is widely used as a model in studies on the development of central nervous system's neurons and on the mechanisms by which they establish integrative circuits. Extensive research on the proliferation, migration, and differentiation of the neurons as well as on the innervation of their targets has already been carried out (Sauer 1935; Bergquist 1964; Nornes and Das 1974; Altman and Bayer 1984), and the [^{3}H]thymidine autoradiographic studies of Nornes and Das (1974) and Altman and Bayer (1984), in particular, have provided a vast amount of results on the development of the neurons of the spinal cord.

In short, the development of the central nervous system (spinal cord) can be considered as the result of several events which overlap in time. Following their proliferation in the matrix layer, the neuroblasts (the mitotic future neurons) migrate towards the periphery to form the mantle layer (Sauer 1935; Hamburger 1975; Bergquist 1964; Cowan 1978). The direction of migration of the neurons is orthogonal to the ventricular surface and most likely closely associated with the radially orientated glial processes (Sidman and Rakic 1973). The direction of movement also seems to be dictated by chemical gradients and/or differential adhesion to the environment (Rakic 1985; Edelman 1985). In the mantle layer, the postmitotic neurons aggregate in anlagen (Rutishauser et al. 1978) and start to differentiate by developing their characteristic morphological, physiological, and chemical properties (Domesick and Morrest 1977; Llinás and Sugimori 1979; Spitzer 1981, 1982; Purves and Lichtman 1985). This phase includes the outgrowth of their axon and dendrites (Van der Loos 1965; Purves and Hume 1981; Bray 1982). Their axonal extension makes contact through synaptic structures on the proper target cell (Sotelo and Changeux 1974; Changeux and Danchin 1976; Sotelo 1981; Changeux 1987). Finally, integrative circuits are determined by adjustments in the already developed connection patterns through elimination of axon collaterals (pruning) and/or entire neurons (neuronal death) or by rearrangement of the connections (Hamburger 1975; Oppenheim et al. 1978; Chu-Wang and Oppenheim 1978a,b; Purves and Lichtman 1980). The availability of specific growth factors seems to play a critical role in this last phase of neuronal development (see Sect. 5 for a closer look at the role of growth factors in spinal cord development).

It should be noted that the development of the central nervous system does not proceed exclusively according to genetically predetermined patterns. Apart from the genetic program that is generally present, the transposition of the linear sequence of genes into a three-dimensional nervous system also requires neuronal competition for environmental epigenetic factors (see Purves and Lichtman 1985; Chandebois 1976). Clearly, the maximum capacity of the genome is inadequate to account for the large variety of cell individualities and neuronal connections that occur in the vertebrate brain. Therefore, proper development can only be accomplished by a diversity of (epigenetic) factors that supplement the genetic information (Chandebois and Faber 1983). One has to bear in mind that these factors do not dictate the developmental programs, but merely reinitiate or deflect their course. The developmental programs, which are established around the time of cell determination, lay down the rules for the exchange of information among the cells, but do not fix their destiny (Chandebois and Faber 1983).

Thus, the development of the central nervous system seems to be determined by the cooperation between the genome and epigenetic factors. Besides this cooperation, development also appears to be dependent on the reciprocal influence between these two factors. In this context, it is important to understand that genetic processes, which seem to act independently of any environmental influence, are actually highly susceptible to shaping by external stimuli (see Chandebois and Faber 1983). Therefore, at the cellular level, the development of the central nervous system is regulated by (genetically cued) intrinsic processes as well as extrinsic processes (the epigenetic factors). Understanding the molecular processes that regulate neural development is currently one of the main challenges in neuroscience. In particular, the isolation and characterization of the genes that encode transcripts for regulators, growth factors, and cell adhesion molecules will contribute to our knowledge about the development of the central nervous system. Because of their explicit role in the establishment of neural and neuronal morphology, the epigenetic factors are usually referred to as morphoregulators.

The extrapolation of the above-mentioned developmental processes to the spinal cord makes it clear that, here too, epigenetic factors are of crucial importance in regulation. Although descriptive or topographical aspects of the developmental processes in the spinal cord are relatively well known, our knowledge about their precise regulation is limited. For instance, much confusion exists about the mechanism of cell recognition, which is considered an important regulator of neural development. Although the first reports on neuronal recognition date back to early this century, the molecular basis is still obscure and subject to extensive debate. Through the years, several hypotheses have been proposed. Among the first was the resonance hypothesis of Weiss (1924), which stated that neural specificity arises through target responsiveness to only those axons that carry appropriate activity patterns ("functional matching"). Also proposed by Weiss (1934) was the contact guidance hypothesis, which contended that axons maintain a relationship with their neighboring axons on the way to their target, where they then connect appropriately because of continued alignment ("mechanical matching"). Though widely rejected by neuroembryologists, the contact guidance hypothesis still appears sporadically in the literature (Horder and Martin 1979).

The chemoaffinity hypothesis, the concept which had already been suggested by Ramon Y Cajal in 1892, but later formulated by Sperry, contends that nerve cells recognize each other by complementary structures present on the cell surface (Sperry 1941, 1963). This attractive idea became generally accepted after the discovery of cell adhesion molecules (CAM; Edelman 1983). Nevertheless, the several different hypotheses are still widely debated. The unraveling of their biochemical basis is an intriguing problem in modern neuroembryology.

The objective of the present study is to describe the appearance and distribution patterns of several putative morphoregulators in the developing rat spinal cord. In combination with the results of intrauterine and postnatal neuronal tracer studies on the development of its fiber systems, an update on the development of the rat spinal cord is presented. In addition to autoradiography, several new techniques, i.e., immunocytochemistry, prenatal tracing, and in situ hybridization, have been successfully applied in neuroembryonic studies during the last few decades. These new techniques have resulted in a more dynamic view of the developmental events in the central nervous system.

The presence of several development-related molecules in the rat spinal cord are reported in this study. In general terms, a detailed knowledge of regional neurochemical central nervous system maturation is important, since it provides information concerning the key events in brain development. In the present study, the studied compounds may function as epigenetic factors which regulate and/or influence the morphological processes in cooperation with the genome (see Edelman 1985). The studied molecules represent a small selection of the large group of potential morphoregulators that are active within the developing central nervous system. One group consists of molecules that are incorporated in the intracellular cytoskeleton and, therefore, can be considered of importance in the development of the characteristic neuronal morphology (Matus 1988; Riederer 1990). Most of these *structural* compounds have also been found to be active in other developmental processes. A second group of morphoregulators is associated with the extracellular cell membrane. These molecules may be considered as *functional* compounds, which, as part of the cell membrane, are accessible to other (complementary) molecules of the extracellular environment. They function as recognition structures in cell–cell interactions and/or axonal guidance and are, therefore, of importance in those processes that ultimately determine the gross neural anatomy of the spinal cord.

In the present study, the development of the spinal cord neuronal cytoskeleton has been investigated with immunocytochemistry for (phosphorylated) neurofilaments and for several microtubule-associated proteins (MAP). These molecules are all incorporated in the axonal cytoskeleton and can, therefore, be considered as structural compounds (Lazarides 1980). The phosphorylation of the carboxy-terminal domains of the neurofilament subunits is considered to be one means by which neurofilaments cross-link and stabilize the axonal cytoskeleton (Hirokawa et al. 1984). The expression of most of the MAP seems to be developmentally regulated (Riederer and Matus 1985; Matus 1988). Earlier studies have suggested that the individual subunits of the neurofilaments as well as different MAP are also implicated in several developmental processes (Bernhardt and Matus 1984; Dahl and Bignami 1985; Anderton et al. 1987).

Besides the above-mentioned markers, the development of the axonal cytoskeleton of the spinal cord fiber systems was also studied by means of neuronal tracers. The development of an intrauterine operation technique (De Beer et al. 1989), together with an improved method for prenatal application of neuronal tracers (Lakke and Hinderink 1989), has paved the way for a more accurate description of the embryonal development of the spinal cord fiber tracts.

The spinal cord glial system has been investigated during development with antibodies against vimentin and glial fibrillary acidic protein (GFAP). Both molecules are incorporated in the glial cytoskeleton and may, therefore, be considered as structural compounds (Lazarides 1980; Bignami et al. 1982; Gard and Lazarides 1982). Vimentin has often been implicated in several developmental processes in the rat spinal cord and is known to be involved in DNA replication and recombination, DNA repair, and gene expression (Lee and Page 1984; Geiger 1987; Georgatos and Blodel 1987a,b).

Of the large pool of functional compounds, two have been studied in relation to the development of the spinal cord; stage-specific embryonic antigen-1 (SSEA-1) and acetylcholinesterase (AChE). The former has been recognized as the carbohydrate $3(\alpha)$-fucosyl N-acetyl lactosamine (FAL), also known as Le^x or X-hapten (Gooi et al. 1981) or CD15, which is present in the neuronal glycocalyx. SSEA-1 was found to be expressed during limited periods during neuronal development (Solter and Knowles 1979; Yamamoto et al. 1985); the carbohydrate chain is known to function as a regulating structure in cell–cell contacts (Glick and Santer 1982).

The distribution pattern of AChE in the developing rat spinal cord has also been studied. During certain periods in development, this enzyme may be associated with the cell membrane, although its exact location has still to be determined. AChE has been investigated in view of its putative role as a developmental regulator (Greenfield 1984; Layer et al. 1988). Its precise function as a morphoregulator, however, is still widely debated.

2 A Survey of the Development of the Rat Spinal Cord

Hitherto, the development of spinal cord neurons has mainly been studied with classical histological staining techniques (His 1889; Ramon Y Cajal 1890; Retzius 1898) and with [^{3}H]thymidine autoradiography (Sauer 1959; Sidman et al. 1959; Fujita 1964; Nornes and Das 1974; Altman and Bayer 1984). During the last two decades, different neuronal tracers were used to study the development of the ascending and descending fiber systems in the spinal cord. Until recently, the use of such tracers in mammals appeared to be restricted to the postnatal period because of difficulties in prenatal administration. The development of an intrauterine operation technique (De Beer et al. 1989) and a method for more accurate prenatal tracer application (see Lakke and Hinderink 1989) appear to offer new possibilities for investigating the embryonal development of fiber systems by means of neuronal tracers. So far, the results indicate the applicability of these techniques in developmental studies of the rat spinal cord fiber systems (Lakke and Hinderink 1989; Lakke and Marani 1991; Wessels et al. 1990). This particular tracing technique is used successfully in the present study.

The following survey of the development of the rat spinal cord cells is based to a large extent on [^{3}H]thymidine autoradiographic studies (Nornes and Das 1974; Altman and Bayer 1984). Cells originate in the subventricular region of the matrix layer and, after a proliferation period, migrate towards the periphery of the neural tube to form the mantle layer. During this proliferation period, the pool of future neurons and glial cells is generated in the matrix layer. Within the mantle layer (or gray matter), the neuroblasts differentiate into neurons. The subsequent survey of the development of the different fiber systems of the spinal cord (present in the marginal layer or white matter) relies in particular on postnatal neuronal tracer studies.

2.1 The Matrix Layer

During the gastrula stage of the rat embryo, interactions between specific mesodermal cells (chordomesoderm) and ectodermal cells initiate the formation of the central nervous system. This process of interactive influences is known as primary neural induction (for a review, see Saxén and Toivonen 1962; Jacobson 1979). Under undisturbed conditions, the induced tissue (prospective neuroectoderm or neural plate) becomes the neural tube by a series of morphogenetic events, which collectively are known as neurulation (reviewed by Karfunkel 1974; see Smits-Van Prooije 1988). It was found that during

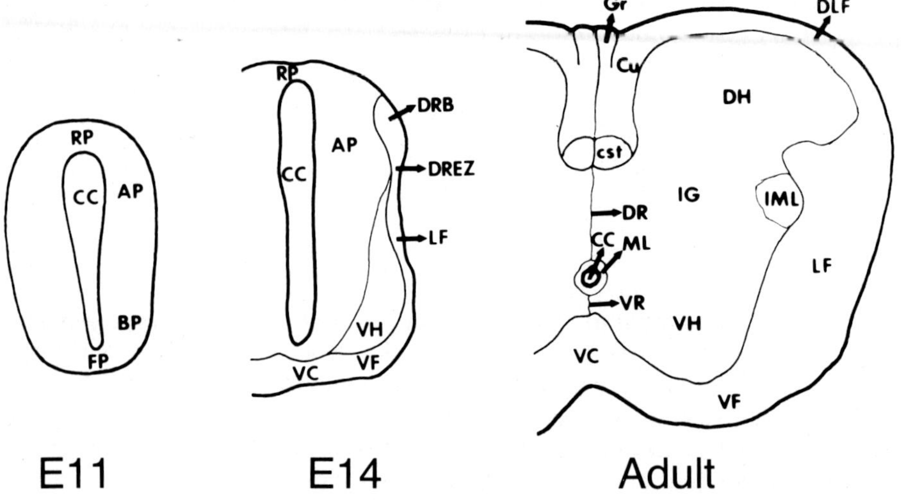

Fig. 2. Transversal cross sections of the developing rat thoracic spinal cord. The drawings illustrate the cord at embryonal day 11 (E11), E14 and at maturity (adult). *RP*, roof plate; *CC*, central canal; *AP*, alar plate; *BP*, basal plate; *FP*, floor plate; *VC*, ventral commissure; *VH*, ventral horn; *VF*, ventral funiculus; *DRB*, dorsal root bifurcation zone; *DREZ*, dorsal root entrance zone; *LF*, lateral funiculus; *cst*, corticospinal tract; *DR*, dorsal raphe; *ML*, matrix layer; *VR*, ventral raphe; *IG*, intermediate gray; *DH*, dorsal horn; *DLF*, dorsolateral fasciculus; *Cu*, fasciculus cuneatus; *Gr*, fasciculus gracilis; IML, intermediolateral cell column

neurulation, the cells migrate due to regional differences and changes in their shape (Burnside and Jacobson 1968). The cell shape appeared to be determined by microtubules (elongation) and microfilaments (apical constriction; Burnside and Jacobson 1968).

In the neuroepithelium layer or matrix layer of the neural tube, four parts can be distinguished: two thicker lateral plates and two thinner "caps." The caps are usually referred to as the (ventral) floor plate and the (dorsal) roof plate (His 1892a,b; Altman and Bayer 1984; see Fig. 2). The stratified appearance of the lateral plates is caused by the migration patterns of the nuclei of the mitotic neuroblasts (Sauer 1935, 1959; Sidman et al. 1959). The lateral plates are usually divided into a ventral basal plate and a dorsal alar plate. The former represents the generation zone of the spinal cord motor neurons, and the latter represents the generation zone of the spinal cord interneurons. Morphologically, the sulcus limitans indicates the border between these two generation plates; it can be recognized at the surface of the ventricle, although it does not stand out at all levels of the cord (see Altman and Bayer 1984). Earlier, a division of the lateral plates into three zones was proposed by Altman and Bayer (1984). The proposed third zone is situated dorsally from the basal plate and is occupied by the future contralaterally and ipsilaterally projecting (relay) neurons (originally designated as commissural and funicular cells, respectively; Ramon Y Cajal 1909).

The neuroblasts and glioblasts are generated in the subventricular zone of the matrix layer, which in general is known to contain the main population of central nervous system stem cells (Hommes and Leblond 1967). In general, the neuroblasts differentiate into the three spinal cord neuronal cell types; motor neurons, relay neurons, and interneurons. For a long time, it was believed that both neuroblasts and glioblasts arise from a common progenitor. However, it has been demonstrated that glioblasts actually derive from a different ancestor, the so-called O2A progenitor cell (Raff et al. 1983). This progenitor cell develops under the influence of the type 1 astrocyte into either a type 2 astrocyte or a oligodendrocyte (Raff et al. 1984). In addition to these glial cell types, microglial cells also arise from the O2A progenitor cell. The timing and direction of the development of the O2A progenitor were found to be strongly influenced by environmental factors (Raff et al. 1984). On the other hand, the development of the three glial cell types appeared to be independent of brain morphogenesis (Abney et al. 1981; Williams et al. 1985). Only recently, the existence of a second glial ancestor has been demonstrated, the so-called MARP (migratory astrocyte-restricted precursor) cell, which was found to give rise to both types of astrocytes (Warf et al. 1991).

The future spinal cord motor neurons are generated in the ventral basal plate (see Fig. 2) between embryonal days 10–14 (E10–E14)[1]. The bulk of them are produced in a 2-day period, E11–E12 (Nornes and Das 1974; Altman and Bayer 1984). The cells migrate in an orthogonal direction towards the periphery of the ventral neural tube (Altman and Bayer 1984) and form the ventral horn of the adult rat spinal cord. In general, chemical gradients, differential adhesion, and radially orientated glial processes are thought to be closely involved in the migration process of the neuroblasts (Sidman and Rakic 1973; Edelman 1985; Rakic 1985). The preganglionic motor neurons of the spinal cord also derive from cells in the ventral part of the matrix layer. These cells are generated roughly between E11 and E13 and later assemble in the intermediolateral nucleus, which forms the lateral horn of the thoracic spinal cord (see Fig. 2).

The ipsilaterally and contralaterally projecting neurons are generated between E11 and E14[2], and the peak production takes place around E12–E13

---

[1] It should be noted that a different time scale to determine the age of the rat embryos was used by Altman and Bayer (1984). In their study, the day of conception was indicated as embryonal day 1, whereas in the present experiments this day was indicated as embryonal day 0. Consequently, in this study the first embryonal day and each day thereafter is set 24 ($\pm$ 3) h later as the numerical corresponding days in the study of Altman and Bayer (1984). For the convenience of the reader, all ages or time periods mentioned in the present study are calculated with the day of conception as embryonal day 0. Those ages that refer to the study of Altman and Bayer (1984) have also been converted according this method.

[2] In some cases the comparison of the described sections as used in the present study with the photographs of the sections as used in the study of Altman and Bayer (1984) seem not to correlate. For instance, the distinct ventral horn in the (Altman and Bayer) E13 rat was not found to be present in our E13 rat. This inconsistency could be caused by the use of different rat strains; the Paurdue Wistar strain (Altman and Bayer) and the Wistar Albino Glaxo strain (present study). Such a difference between rat strains has been found before (see Angulo Y González 1940 and Altman and Bayer 1984). To avoid any confusion, the time scale as described in footnote 1 (see also Material and Methods and footnote 3) was applied in all cases, regardless of differences in the appearance of the spinal cord in the Alman and Bayer study (1984).

(Altman and Bayer 1984). The contralaterally projecting neurons are generated before the ipsilaterally projecting neurons (Altman and Bayer 1984). The cells migrate towards the periphery of the intermediate region of the neural tube and form the intermediate gray of the adult rat spinal cord (Nornes and Das 1974; Altman and Bayer 1984). Recently, prelabeling experiments with the fluorescent neuronal tracer DiI demonstrated that at least seven different subpopulations of future intermediate gray cells can be recognized in the rat spinal cord matrix layer. These groups of cells appear to originate at different places along the spinal cord neuroepithelium, adjacent to their future location in the mantle layer (Silos-Santiago and Snider 1990).

The production of the future interneurons takes place in the dorsal alar plate (see Fig. 2) and lasts from E13 to E16; their peak production is around E14–E15 (Altman and Bayer 1984). The cells of the alar plate migrate towards the periphery of the dorsal neural tube and later form the dorsal horn of the adult rat spinal cord (see Fig. 2; Nornes and Das 1974; Altman and Bayer 1984).

For a long time, the functions of the floor and roof plate were obscure. They are thought to be the source of neuroglial cells such as crest cells and fibrous glia. The crest cells give rise to the peripheral Schwann cells, whereas the fibrous glia forms a component of the permanent ependymal lining (see Altman and Bayer 1984). Nevertheless, conclusive evidence for such a role is lacking. A second and more plausible suggestion is that the resilience of radial glial fibers, which were found to have their origin in the floor and roof plate, leads to the formation of the spinal cord ventral and dorsal midline structures (see Altman and Bayer 1984). The midline structures could function as axon barriers, which regulate the course of the commissural and dorsal column fibers. Although several axon guidance molecules were found to be present in the roof plate, a direct relation to the proposed barrier function is still uncertain (Snow et al. 1990).

Only recently did our knowledge about the floor plate begin to increase. It was demonstrated that the floor plate is the source of inductive signals, which are believed to attract commissural axons to the ventral midline, whereas other axons may be inhibited from crossing to the contralaterally side of the cord (Dodd and Jessel 1988; Tessier-Lavigne et al. 1988). The floor plate also seems to induce the development of the motor neurons of the ventral mantle layer (Tessier-Lavigne et al. 1988; Glover 1991).

2.2 The Mantle Layer

The mantle layer is formed by the neuroblasts that migrate towards the periphery of the matrix layer after they have ceased their mitotic cycles. The future neurons appear to migrate in the direction of their appropriate location in the spinal cord. A clear topographical relation was found between the matrix layer and the mantle layer; the neuroblasts of the ventral matrix layer settle in the ventral mantle layer (Nornes and Das 1974). This was also found to pertain for the dorsal regions of the two layers. Additionally, Altman and Bayer (1984) proposed such a topographical relation for the intermediate regions after they demonstrated that neurons of the intermediate gray area originated in an intermediate zone of the matrix layer, situated dorsally from the ventral basal plate. So far, additional conclusive evidence for such a third generation zone has not been provided. In contrast, a recent study demonstrated that different groups of future intermediate gray cells develop in the matrix layer opposite their future location in the mantle layer (Silos-Santiago and Snider 1990; see above). This suggests the presence of a large number of neuroblast subpopulations in the matrix layer, although it also warrants a general topographical relation between the matrix and mantle layer of the rat spinal cord.

The ventral part of the mantle layer is occupied by the spinal cord motor neurons and is known as the ventral horn (Fig. 2). An incipient ventral horn can already be distinguished at E12 (Altman and Bayer 1984). During the following days, the ventral neuroblasts differentiate into the spinal cord motor neurons. Five different phases can be recognized in this process. The first three phases (axonogenesis) are characterized by a rounding up and a lateral movement of the neuroblasts, the formation of a laterally orientated basal cytoplasmatic cap, and the outgrowth of the axon from this cap (Altman and Bayer 1984). The last two phases (dendrogenesis) are characterized by the development of first a bipolar and later a multipolar appearance of the motor neurons and the outgrowth of its dendrites (Altman and Bayer 1984). Finally, three types of motor neurons can be distinguished; the larger α-motor cells and the smaller β- and γ-motor cells. The α-motor neurons appear to be generated before the β- and γ-neurons (Nornes and Das 1974). The α-motor cells were also found to be the earliest cells to mature in the spinal cord (Altman and Bayer 1984). At the level of the cervical and lumbar enlargements, the motor neurons are aggregated in the lateral and medial motor columns that innervate the skeletal muscles of the body wall and extremities and the axial muscles, respectively (Goehring 1928; Sprague 1958; Sterling and Kuypers 1967; Hollyday 1985). The assembly of the motor neurons into columns starts around E13 and is completed around E17 (Altman and Bayer 1984). In the adult rat spinal cord, the motor columns are usually referred to as Rexed's lamina IX (Rexed 1954). At the thoracic level of the spinal cord, only a single motor column can be found. At this particular level, the lateral horn of the spinal cord mantle layer is formed by the intermediolateral nucleus, which contains the preganglionic motor neurons (Rando et al. 1981; Altman and Bayer 1984). These cells are generated between E11 and E13, which is the same period in which the thoracic motor neurons are produced (Altman and Bayer 1984). The lateral horn is the most lateral part of Rexed's lamina VII in the adult rat spinal cord (Rexed 1954). At lumbar levels, preganglionic motor neurons are also located in the intermediolateral nucleus, although a clear lateral horn cannot be distinguished (Schramm et al. 1975; Rando et al. 1981).

The intermediate part of the mantle layer, which later develops into the intermediate gray (see Fig. 2), is occupied by the contralaterally or ipsilaterally projecting neurons. These relay cells are generated during a 3-day period, E11–E13, with a peak production at E12 (Altman and Bayer 1984). The relay neurons appear in a wide range of sizes ($8-40\,\mu$m) and shapes (spindle-, star-, and triangular-shaped; see Brown 1981). The controlaterally projecting cells are generated before the ipsilaterally projecting cells (Altman and Bayer 1984).

The neurons of the intermediate gray settle in Rexed's laminae I and IV–VIII of the adult rat spinal cord (Rexed 1954). Their axons innervate local spinal levels (propriospinal axons; Kuhlenbeck 1975; Brown 1981) or higher brain centers (Molenaar and Kuypers 1978; Menétrey et al. 1982). The incipient intermediate gray can be distinguished around E13 (Altman and Bayer 1984). During further development, the most laterally situated relay neurons seem to be pushed by mechanical actions to the most dorsal layer of the dorsal horn (see Altman and Bayer 1984). This layer is known as Waldeyer's layer or the marginal layer and forms Rexed's lamina I in the adult rat spinal cord (Rexed 1954).

The dorsal part of the spinal cord mantle layer is known as the dorsal horn (Fig. 2) and is occupied for the most part by interneurons. A small part, Rexed's lamina I, is occupied by the marginal cells of Waldeyer's layer (see above). The dorsal horn contains a heterogeneous group of cells of which the smaller ones are situated in the substantia gelatinosa or Rexed's lamina II, whereas the larger ones are located in Rexed's lamina III. The interneurons were found to project into the spinal cord level in which they are situated (the reflex circuit) or to higher brain areas (Burton and Loewry 1976). The formation of the dorsal horn, Rexed's laminae I–III in the adult rat spinal cord, starts around E13–E14 (Altman and Bayer 1984).

From a developmental point of view, two patterns appeared to be present in the formation of the spinal cord mantle layer. Firstly, a ventral-to-dorsal gradient in the generation of the different cell classes (motor and relay neurons and interneurons) in the matrix layer. The migration of these cells from the inner subventricular area towards the periphery of the matrix layer also follows a ventral-to-dorsal gradient. The second developmental pattern is a rostral-to-caudal gradient, which can be most clearly discerned in the cervical spinal cord. (Nornes and Das 1974; Altman and Bayer 1984).

2.3 The Marginal Layer

The development of the marginal layer of the rat spinal cord depends on the outgrowth of the axons of spinal cord neurons and of the arrival of the axons of neurons in higher brain regions and in the periphery. Specific developmental studies on the fiber systems in the fetal rat spinal cord are limited, since prenatal neuronal tracing has only recently become fruitful in mammals (see Lakke and Hinderink 1989; Wessels et al. 1990). The following description of the early, embryonic development of the spinal cord fiber systems is, therefore, mainly based on the results of (^{3}H)thymidine autoradiographic studies (Sauer 1959; Sidman et al. 1959; Fujita 1964; Nornes and Das 1974; Altman and Bayer 1984) and some classical silver impregnation studies (Ramon Y Cajal 1909; Bodian 1936; Windle and Austin 1936). The description of the late, postnatal development of the spinal cord fiber tracts is merely based on postnatal tracer studies.

The first ventral motor roots leave the spinal cord around E12. One day later, the central processes of the dorsal ganglia cells, assembled in the dorsal roots, were found to enter the cord at the dorsal root entrance zone and to form the oval-shaped dorsal root bifurcation zone (see Fig. 2; Altman and Bayer 1984; Oudega et al. 1992a). On this day, E13, the first contralaterally projecting fibers can be recognized in the ventral commissure and the ventral funiculus. On the same day, although significantly less in number, ipsilaterally projecting axons could also be found in an incipient lateral funiculus (Oudega et al. 1992a). Between E14 and E17, the dorsal root bifurcation zone develops into the dorsal funiculus, which contains the large myelinated ascending propriospinal fibres. The first ascending fibers in the dorsolateral fasciculus can be discerned around E16 (Altman and Bayer 1984; Oudega et al. 1992a).

Unfortunately, the usefulness of silver impregnation methods in ontogenetic studies is restricted to the early developmental period, because during later periods an accurate discrimination between fiber tracts cannot be guaranteed. Nevertheless, Ramon Y Cajal was able to identify a descending pathway in the early development of the spinal cord by means of silver staining. These fibers coursed along the ventral funiculus and belong to the reticulospinal tract, which is part of the fasciculus longitudinalis medialis (Ramon Y Cajal 1909). For the identification of individual ascending and descending fiber systems in the developing mammalian spinal cord, the use of neuronal tracers seems unavoidable, and the regional application of tracers can produce a more specified and dynamic view of the development of the fiber systems. Consequently, the early development of many of the spinal cord fiber systems makes prenatal use of the tracer a necessity. In the present study, prenatal tracing in mammals will be described, and its results and applicability will be discussed.

The following survey on the development of the spinal cord fiber tracts deals with postnatal tracer studies in different animal species. An extrapolation of these results to the rat seems warranted by the observation that the ontogeny of most of the spinal cord fiber systems is similar in a wide variety of animals (Ten Donkelaar 1982; Martin et al. 1978, 1982). This summary does not intend to be complete; in fact, only the major fiber systems will be mentioned (for a recent review on the adult rat spinal cord fiber systems, see Tracey 1985).

The ventral funiculus of the rat spinal cord is primarily occupied by the descending fiber tracts of the fasciculus longitudinalis medialis, which contains the crossed tectospinal, crossed and uncrossed reticulospinal, uncrossed coeruleospinal, crossed and uncrossed vestibulospinal, and crossed and uncrossed interstitiospinal fiber systems (Waldron and Gwyn 1969; Giesler et al. 1979; Menétrey et al. 1982). Most of these tracts were found to develop well before birth (Martin et al. 1978, 1982, 1983; Okado and Oppenheim 1985). The earliest recognizable fiber system in the rat spinal cord is the reticulospinal tract, something which was already known from early silver stain studies (Ramon Y Cajal 1909; Windle and Austin 1936).

The lateral funiculus contains ascending as well as descending fiber systems (Zemlan et al. 1978; Giesler et al. 1979, 1981; Ruggiero et al. 1981; Menétrey et al. 1982). Among the many fiber tracts are the ascending uncrossed dorsal and crossed ventral spinocerebellar tract (Oscarsson 1973; Zemlan et al. 1978), of which the first fibers reach the cerebellum before birth (Arsénio-Nunes and Sotelo 1985). Another extensively studied system of the lateral funiculus is the descending crossed rubrospinal fiber tract. The first rubrospinal fibers reach the higher cervical levels of the spinal cord well before birth, whereas the lower lumbar levels appeared to be reached around the day of birth (Cabana and Martin 1982, 1986).

The dorsal funiculus contains the ascending, uncrossed, thick, myelinated branches of the larger dorsal ganglia cells (Basbaum and Hand 1973; Burton and Loewry 1976) as well as the descending fibers of the extensively studied corticospinal tract (Brown 1971; Kuypers 1981; Miller 1987). Recently, it was demonstrated that the dorsal funiculus also contains a small portion of unmyelinated fibers (Patterson et al. 1989). The rat corticospinal tract courses along the ventral part of the contralateral dorsal column. Around birth, corticospinal fibers appear to be present in the higher cervical cord levels. During the second postnatal week, the lumbar and sacral levels are reached (Gribnau et al. 1986; Joosten et al. 1987).

After birth, oligodendrocytes start the myelination process within the spinal cord, and during the first 3 weeks, all fiber tracts were found to be myelinated (Matthews and Duncan 1971; Hirano and Dembitzer 1978; Schwab and Schnell 1989). The myelin sheaths around the axons give the marginal layer its white, opaque appearance and its common name "white matter." Surprisingly, the myelination of both the descending and the ascending fiber tracts was found to proceed according to a rostral-to-caudal gradient (Schwab and Schnell 1989). Recent studies have shown that the central nervous system myelin contains at least two neurite growth inhibitors (Schwab 1990a,b). Therefore, in the last stage of spinal cord development, the increase of the myelin content could account for an inhibitory effect on the growth of the fiber systems (Schwab 1990a,b; Schwab and Schnell 1991). In addition, the inhibitory properties of myelin could also function as a boundary or as a guidance factor for the "late"-growing spinal cord fiber tracts (Schwab and Schnell 1991).

2.4 Development of Descending Fiber Tracts

Suprasegmental descending projections (SDP) take their time to develop. 5-Hydroxy-tryptamine-immunoreactive fibers originating in the medullary raphe nuclei enter the cervical spinal cord at E14 (?)[2], and reach the sacral spinal cord at E18 (?) (Rajaofetra et al. 1989). The first corticospinal axons enter the spinal cord between E21 (?) and postnatal day 1 (P1) and reach the sacral

---

[2] Several systems are currently used to number gestational days (see also footnote 1). We have adopted the system recommended by Paxinos et al. (1991), which designates the day of insemination as embryonic day 0 (E0) and likewise the day of birth as postnatal day 0 (P0). In all tracer experiments, rat dames in heat (discriminated by their behavior; Zwet et al. 1986) were separated from the colony and bred with an active rat bull for an 1-h period. The end of this period was designated as the beginning of E0. As far as possible all the cited indications of gestational age calculated in different systems have been converted to the Paxinos system. Converted values are indicated by parentheses. If no system was explicitly stated, a question mark in parentheses follows. If nothing is indicated, our dating and the cited dating coincide.

spinal cord at P10 (Gribnau et al. 1986; Schreyer and Jones 1982). In both cases, a delay exists between the arrival of fibers (in the white matter) at a certain level of the spinal cord and their invasion of the gray matter of the same cord level: 1 day in the case of the raphe-spinal projection and 2 days in the case of the corticospinal projection. A timetable of the descent and ingrowth of other SDP will reveal the significance of this delay difference and might yield insights into factors regulating the development of these complex projection systems, for instance, by comparing it to the cytogenetic chronology (of source and target), distance to target, projection mode (diffuse, circumscribed, somatotopic, etc.), or transmitter content. Furthermore, such data will aid us in the interpretation of data on the appearance, distribution, and disappearance of various morphoregulators and chemodeterminants during the development of the spinal cord.

To construct such a timetable, the development of individual SDP has to be traced through their development. Some SDP distinguish themselves by the expression of a specific marker and can be described by detection of this marker (Bregman 1987; Bernstein-Goral and Bohn 1988; Rajaofetra et al. 1989). SDP lacking such a marker can be described by conventional anterograde or retrograde axonal tracing (Distel and Hollander 1980; Okado and Oppenheim 1985; Lakke and Hinderink 1989; Lakke and Marani 1991).

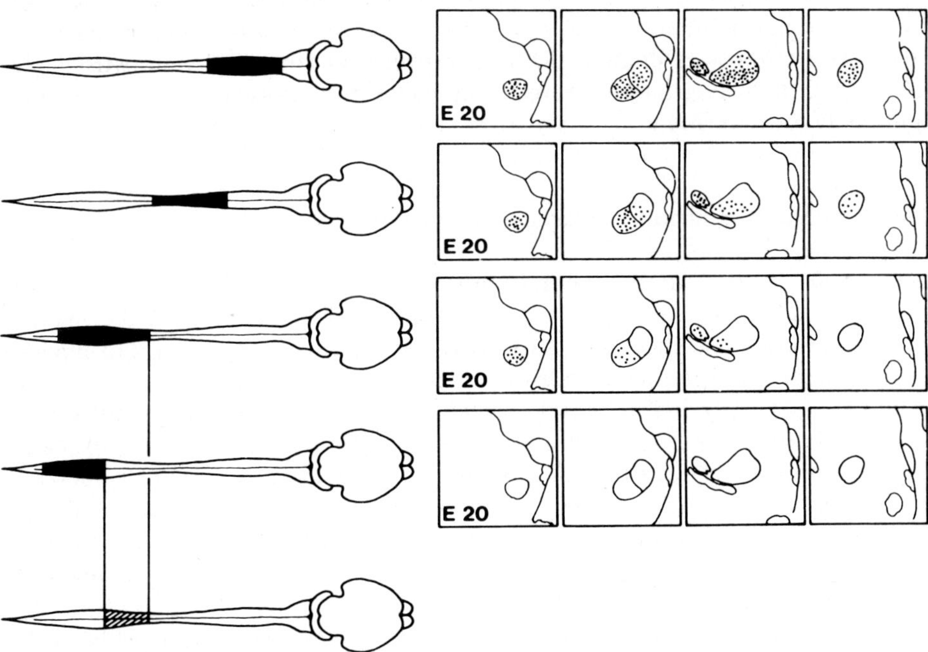

Fig. 3. Construction of a (fictive) position interval of descending fibers. The descending fiber tip is located between the rostralmost border of the caudalmost injection site which *did* result in retrograde labeling of the pertinent nucleus and the rostralmost border of the rostralmost injection which did *not* result in labeling of the pertinent nucleus. In those cases in which no "negative" experiment was available, the caudal end of the spinal cord was used as the caudal border of the position interval (see Table 4). *E20*, embryonal day 20

Anterograde tracing theoretically offers the highest location resolution in the spinal cord, but each SDP source nucleus (SDPsn) has to be investigated separately, necessitating high-precision injection techniques in fetal and neonatal rats. Retrograde tracing investigates all SDP at the same time, allowing a substantial reduction in the number of experiments necessary.

Retrograde neuronal tracers (horseradish peroxide, HRP, and wheat germ agglutinin, WGA–HRP) were injected into the spinal cord of rat fetuses and neonates, both at different gestational ages and at different levels of the spinal cord. By comparing the level of the injection sites of each age group to the retrograde labeling in a certain SDPsn, a position interval can be reduced for the leading descending fibers of this SDP (Fig. 3); this interval is located between the rostral border of the caudalmost injection which *did* result in retrograde labeling in the pertinent SPDsn and the rostral border of the rostralmost injection which did *not* result in retrograde labeling of this SDPsn (Lakke and Hinderink 1989; Lakke et al. 1990; Lakke and Marani 1991). In the case of a larger SDPsn, an axonogenetical gradient can be deduced (Lakke and Marani 1991). The descending fiber systems described in this paper were selected according to the position of the fibers in the spinal cord funiculi: one in the ventral funiculus (coeruleospinal fibers), one superficially in the dorsolateral funiculus (spinal projection from the nucleus tegmentalis laterodorsalis, TLD), and one deep in the dorsolateral funiculus (rubrospinal projection). In the rat, the corticospinal tract is located deep in the dorsal funiculus (Brown 1971). The developmental descent of the corticospinal tract has been described by several authors (Donatelle 1977; Schreyer and Jones 1982; Joosten et al. 1987), and their descriptions serve to supplement the present data with data on a dorsally located SDP.

2.4.1 Nucleus Tegmentalis Laterodorsalis: Dorsolateral Funiculus

The TLD, located in the lateral part of the pontine tegmentum, is bordered at its caudal end by the locus coeruleus (LC) laterally and the nucleus tegmentalis dorsalis medially. Rostrally, the TLD lies between the nucleus raphe dorsalis and the pedunculus cerebellaris superior (Tohyama et al. 1978; Andrezik and Beitz 1985). Spinal projections from the TLD actually arise from a TLD subnucleus, Barrington's nucleus (Tohyama et al. 1978; Satoh et al. 1978b), located between the LC caudally and the TLD proper rostromedially (Tohyama et al. 1978). In the spinal cord, the descending TLD fibers are located superficially in the lateral funiculus (Loewy et al. 1979; Rye et al. 1988) and project to the intermediolateral cell column of the sacral spinal cord (Loewy et al. 1979; Saper and Loewy 1980).

Micturition in the adult rat is mediated by a spinobulbospinal reflex pathway, which is activated by bladder distention. The bulbar micturition reflex center is located in the pontine tegmentum (Barrington 1925; Satoh et al. 1978a; Loewy et al. 1979; Kruse et al. 1990). In the cat, this reflex center consists of at least two regions, i.e., the M-region, located in the medial dorsolateral pontine tegmentum, and the L-region, located in the lateral pontine tegmentum (Holstege et al. 1986). Stimulation of the M-region elicits a decrease in urethral sphincter pressure followed by an increase in intravesical pressure (Holstege et al. 1986; Holstege and Griffiths 1990). In the rat, similar effects are obtained through the stimulation of either the TLD or the nucleus parabrachialis lateralis (Satoh et al. 1978a; Kruse et al. 1990). Of these two nuclei, only the TLD projects directly onto the sacral parasympathetic moto-neurons (Satoh et al. 1978b; Loewy et al. 1979; Saper and Loewy 1980).

In the neonatal rat, micturition is initiated by the mother licking the perineal region, thus activating a spinal reflex. However, a weak spinobulbo-spinal reflex is present from at least P2 (Kruse and de Groat 1990), indicating that the spinal projection of the TLD has developed functional synapses. A preliminary report on the development of the SDP from the TLD has been published (Lakke et al. 1990).

2.4.2 Locus Coeruleus; Ventral Funiculus

The LC is located in the pontine tegmentum near the lateral edge of the floor of the rostral part of the fourth ventricle. The LC extends from a level just rostral to the nucleus princeps V up to the level of the TLD. Caudally, the LC lies lateral, rostrally rather ventrolateral to the central pontine gray. Rostro-ventrally to the LC the nucleus subcoeruleus (SC) is located. The LC cor-responds loosely to the noradrenergic cell group A6 of Dahlström and Fuxe (1964). Apart from noradrenaline, some LC neurons contain vasopressin, neurophysin (Caffe and van Leeuwen 1983), neurotensin (Uhl et al. 1979), corticotropin-releasing factor (Swanson et al. 1983), neuropeptide Y, and galanin (Holets et al. 1988). The LC consists of a compact dorsal and a more loosely packed ventral part (Swanson 1976).

Fibers from the LC descend through the ipsilateral tractus tegmentalis centralis and through the ipsilateral ventral funiculus to the lowermost levels of the spinal cord (Jones and Yang 1985; Clark and Proudfit 1991). Some authors claim that the noradrenergic fibers from the LC descend almost solely through the superficial laminia of the dorsal horn (Fritschy et al. 1987; Fritschy and Grzanna 1990a). Within the brainstem, the descending coerulean fibers ter-minate mainly in the somatic afferent nuclei, the pontine nuclei, the nucleus interpeduncularis, and within the nucleus oliva inferior (Westlund et al. 1981, 1983; Jones and Yang 1985; Fritschy and Grzanna 1990b). Within the spinal cord, fibers from the LC terminate diffusely in Rexed's laminae IV-X at all levels of the cord (Jones and Yang 1985; Clark and Proudfit 1991). Again, some authors claim a more localized projection of noradrenergic fibers from the LC to Rexed's laminae I, II, and X only (Fritschy et al. 1987; Fritschy and Grzanna 1990a). LC fibers descending to the spinal cord originate mainly from the ventral and caudal LC (Satoh et al. 1977; Westlund et al. 1983; Clark and Proudfit 1991).

Most (95%) of the LC neurons differentiate from the ventricular matrix during E12, and the remainder during the following 2 days; the last neurons generated are deposited dorsomedially in the LC, suggesting a ventrolateral-to-dorsomedial generation gradient (Altman and Bayer 1980d). Synthesis of the monoaminergic neurotransmitter norepinephrine was first demonstrated shortly after the termination of the generation period (Lauder an Bloom 1974). At birth, both ventrally and dorsally located LC neurons project to the spinal cord (Chen and Stanfield 1987). During the following 4 weeks, the dorsally located neurons loose their descending (spinal) collateral, apparently in favour of an ascending collateral (Chen and Stanfield 1987; Stanfield 1989).

2.4.3 Nucleus Ruber; Dorsolateral Funiculus

The nucleus ruber (NR) of the rat is located in the rostromedial mesencephalic tegmentum. The NR extends from the plane of the oculomotor nerve caudally to the plane of the fasciculus retroflexus rostrally. The NR is subdivided into a rostral, parvocellular (pcNR) and a caudal, magnocellular (mcNR) portion (Gillilan 1943; Reid et al. 1975). Within the mcNR, a caudal pole (cNR) and a dorsomedial (dmNR) and ventrolateral (vlNR) subgroup are discerned. A lateral protrusion extends from the ventrolateral aspect of the mcNR.

Through the rubrobulbar and rubrospinal tracts, the NR projects contralaterally to the nucleus reticularis lateralis, the nucleus facialis, the nucleus interpositus, the nucleus vestibularis descendens, the nucleus cuneatus, and the spinal cord and ipsilaterally to the nucleus ventralis lateralis thalami and the nucleus olivaris inferior (Held 1890; Waldron and Gwyn 1969; Flumerfelt and Gwyn 1974; Brown 1974b; Murray and Gurule 1979; Hinrichsen and Watson 1983; Daniel et al. 1987; Kennedy 1987).

The rubrospinal projection originates in large and small neurons located throughout the NR, though mainly from neurons in the caudal two thirds of the nucleus. This projection is organized somatotopically; the vlNR projects to the lumbosacral spinal cord, and the dmNR to the cervical spinal cord (Brown 1974b; Flumerfelt and Gwyn 1974; Reid et al. 1975; Murray and Gurule 1979; Huisman et al. 1981; Shieh et al. 1983; Daniel et al. 1987).

The rubrospinal tract emerges from the ventromedial aspect of the NR and decussates in the ventral tegmentum. It courses through the ventrolateral part of the brainstem, passes into the dorsolateral funiculus, and descends all the way down the spinal cord. The rubrospinal projection terminates contralaterally, in the lateral half of Rexed's lamina V and in Rexed's lamina VI, at all levels of the spinal cord (Waldron and Gwyn 1969; Brown 1974b; McClung and Castro 1978).

At P3, the rubrospinal tract of the rat has reached the lumbosacral spinal cord, and the rubrospinal somatotopy has already become established (Shieh et al. 1983; Leong et al. 1984). A report on the development of the rubrospinal tract has already appeared (Lakke and Marani 1991).

3 Materials and Methods

3.1 Animals

Adult Wistar albino Glaxo rats were housed under standard conditions (light/dark regime; light period 8 A.M.–4 P.M., relative humidity, 50%–60%). Mating was allowed during a 4-h period, from 8–12 A.M. After anesthesia of the timed-pregnant mother with Hypnorm (0.3 ml/kg, s.c.) and Valium (diazepam, 1 ml/kg, i.m.), the fetuses were taken from the uterus. The embryos were routinely removed between 9 and 11 A.M. Consequently, the age of the fetuses could be determined within a margin of error of about 3 h (van der Zwet et al. 1986). The first 24-h period after mating was considered to be day 0 of development (E0). Postnatal animals were anesthetized with Valium (1 μl/g, s.c.) and Hypnorm (0.3 μl/g, i.m.) before being further histologically processed. The day of birth, usually E22, was indicated as postnatal day 0 (P0). The postnatal rats were taken from different litters. A total of 481 animals of 26 different ages were studied. In the embryonal life span, all ages from E9 to E22 were examined. Each age was represented by at least 22 embryos. The postnatal groups comprised a minimum of 15 rats. The examined postnatal ages were: P1, 2, 4, 6, 8, 10, 12, 16, 20, 24, 30, and maturity.

3.2 Histological Procedures

The spinal cord of the animals used or, in the case of the younger embryonal ages, the whole animal was processed for nonfixed cryostat series or fixed gelatin series. The cryostat series were used for both immunohistochemical and the enzyme histochemical staining. The gelatin series were exclusively used for enzyme histochemistry. In the cryostat series, each embryonal age was represented by a minimum of 14 animals, whereas each postnatal age comprised at least eight animals. In the gelatin series, the embryonal groups included at least ten animals per age, whereas the postnatal groups included a minimum of five animals each. In the tracer studies, the animals were processed slightly differently to visualize the injected tracer (see Sect 3.6).

For the cryostat series, nonfixed embryos younger than P0 were placed on a tissue holder immersed in Tissue-Tek O.C.T. compound (Miles Laboratories, Naperville, U.S.A.). The embryos were frozen by placing the tissue holder in isopentane chilled with liquid nitrogen (Marani 1978). This technique results in a gentle freezing of the tissue, which especially benefits the embryonal material, because of its vulnerability to freezing artifacts. Transversal or sagittal sections (10 μm) were mounted on chrome alum–gelatin-coated slides, fixed in acetone (−20°C), and air-dried overnight. Finally, the sections were fixed for 10 min in acetone at room temperature and air-dried. Whole spinal cords of rats older than P0 were dissected out, frozen, fixed, and cryostat-sectioned, as described above. If not used immediately, the sections were stored at −20°C until further histochemical processing.

For the gelatin series, animals younger than E16 were removed from the uterus and fixed by immersion with 4% paraformaldehyde in 0.1 M phosphate buffer (pH 7.4) at 4°C for 48–72 h. Fetuses older than E16 and postnatal animals were transcardially perfused with saline (with 6000 I.U. of heparin per 100 ml) followed by 4% paraformaldehyde in 0.1 M phosphate buffer (pH 7.4). The spinal cord was dissected out and postfixed by immersion with the same fixative for 24 h at 4°C. After washing in buffer, the tissue was embedded in 12% gelatin (Difco Laboratories, Detroit, U.S.A.) in phosphate-buffered saline (PBS, pH 7.4) and hardened for approximately 15 min at -20°C. The gelatin blocks were fixed overnight in the described aldehyde fixative. Transversal sections ($30\,\mu$m) were made with a Jung freeze microtome and selected in a 50 mM TRIS-maleic buffer (pH 7.4).

3.3 Antibodies

In all, six different monoclonal antibodies (MAb) and one polyclonal antibody (PAb) were used in this study. Peroxidase-conjugated rat-anti-mouse antibodies or peroxidase-conjugated goat-anti-rabbit (GAR/PO) antibodies were used in the second phase of the immunohisto-chemical staining procedures. Table 1 presents the antibodies used with some characteristics and references.

The murine MAb NF-90 is directed against neuron-specific neurofilament subunits. NF-90 appears to recognize the phosphorylated form of the three subunits, low molecular weight (NF-L), middle molecular weight (NF-M), and high molecular weight neurofilament protein (NF-H; relative molecular masses 70, 150, and 200 kDa, respectively). The preparation and characterization of this antibody has been described previously (Oudega 1990). The neurofilaments are incorporated in the axonal cytoskeleton and are thought to be involved in the mechanical integration of the organelles in the axoplasmic space (Lazarides 1980). The phosphorylation of the neurofilament subunits is considered as one means by which the neurofilaments cross-link and stabilize the axonal cytoskeleton (Hirokawa et al. 1984; Nixon and Sihag 1992).

The MAb MAP 1A, MAP 2, and MAP 5 were all raised in the mouse and recognize MAP. MAP 1A (clone BW6, kindly donated by Dr. B. Riederer, University of Lausanne, Switzerland) was found to recognize the MAP 1A protein, which has a relative molecular mass of 350 kDa (Riederer et al. 1986, 1991; Garner et al. 1990). This anti-MAP 1A antibody has been described previously (Garner et al. 1990). MAP 2 (clone C, kindly donated by Dr. A.I. Matus, FMI, Basel, Switzerland) was found to recognize the 280-kDa doublet protein bands of MAP 2 (MAP 2A and MAP 2B) as well as the smaller MAP 2C. This antibody has been described earlier (Huber and Matus 1984; Riederer and Matus 1985). MAP 5 (clone AA6) stains a pair of bands with relative molecular mass of 320 kDa and has been described before (Riederer et al. 1986). In general, MAP are thought to be involved in the development (the early MAP) and in the stabilization and organization (the late MAP) of the microtubules (Matus 1988; Riederer 1990). MAP 1A and MAP 2A are considered to be late MAP, whereas MAP 2B, MAP 2C, and MAP 5 are known as early MAP (Binder et al. 1984; Huber and Matus 1984; Bernhardt et al. 1985; Riederer and Matus 1985; Riederer et al. 1986; Tucker and Matus 1988; Tucker et al. 1988; Schoenfeld et al. 1989).

The murine MAb 3B9 is directed against the carbohydrate FAL (also known as Le^x, X-hapten, or CD15), which has been designated SSEA-1 (Solter and Knowles 1979; Fox et al. 1983; Kannagi et al. 1983; Feizi 1985). Antibodies recognizing the trisaccharide FAL are classified as Cluster of Differentiation (CD)15 (International typing workshop, Paris, 1982). All CD15 antibodies recognize the epitope FAL. The antibody 3B9 (also known by its synonym CLB gr/2) was produced and specified by the Central Laboratory of the Netherlands Red Cross Blood Transfusion Service (see Tetteroo et al. 1984). SSEA-1 is implicated in neuronal recognition processes, especially in the sensory system of the spinal cord (Dodd and Jessel 1985, 1986; Jessel and Dodd 1985).

The murine MAb V9 and the rabbit PAb anti-GFAP are directed against vimentin and GFAP, respectively. Both antibodies have already been specified and described – V9 (Sanbio, Uden, the Netherlands) by van Muijen et al. (1984) and anti-GFAP (DAKO no. Z334, Glostrup, Denmark) by Baumal et al. (1980) and Tascos et al. (1982). Vimentin, with a relative molecular mass of 57 kDa, is present in the cytoskeleton of eukaryotic cells of mesenchymal origin (Cabral and Gottesman 1979; Gard and Lazarides 1982). Besides its role in the cytoskeleton (Lazarides 1980; Traub 1985), several other functions have also been suggested for vimentin (Cochard and Paulin 1984; Geiger 1987; Georgatos and Blodel

Table 1. Antibodies used in the present study and some of their characteristics

Antibody	Name/synonym	MAb/PAb	Antigen/relative molecular mass	References
Anti-NF	NF-90	MAb/mouse	Phosphorylated NF-L, NF-M, NF-H (M_r 70, 150, 200 kDa)	Oudega 1990
Anti-MAP 1A	MAP 1A (clone BW6)	MAb/mouse	MAP 1A (M_r 350 kDa)	Riederer et al. 1986 Garner et al. 1990 Oudega et al. 1992
Anti-MAP 2	MAP 2 (clone C)	MAb/mouse	MAP 2A, 2B, 2C (M_r 280, 280, 70 kDa)	Huber and Matus 1984 Riederer and Matus 1985 Oudega et al. 1992
Anti-MAP 5	MAP 5 (clone AA6)	MAb/mouse	MAP 5 (M_r 320 kDa)	Riederer et al. 1986 Oudega et al. 1992
Anti-SSEA-1	3B9/CLB gr/2	MAb/mouse	3(α)-Fucosyl N-acetyl lactosamine (CD15) or SSEA-1	Tetteroo et al. 1984 Oudega et al. 1992
Antivimentin	V9	MAb/mouse	Vimentin (M_r 57 kDa)	Van Muijen et al. 1984 Oudega and Marani 1991
Anti-GFAP		PAb/rabbit	GFAP (M_r 52 kDa)	Baumal et al. 1980 Tascos et al. 1982 Oudega and Marani 1991

MAP, microtubule-associated protein; SSEA, stage-specific embryonic antigen; GFAP, glial fibrillary acidic protein; NF, neurofilament; MAb, monoclonal antibody; PAb, polyclonal antibody; NF-L, low molecular weight NF; NF-M, medium molecular weight NF; NF-H, high molecular weight NF.

1987a,b). GFAP (relative molecular mass, 52 kDa) is present in the cytoskeleton of proto-plasmic and fibrous astroglial cells (Ludwin et al. 1976; Bovalenta et al. 1987). Functions of GFAP other than its role in the cytoskeleton have been suggested (Hansen et al. 1989).

3.4 Immunohistochemistry

The acetone-fixed cryostat sections were rinsed three times for 10 min in PBS (pH 7.4). The third bath contained 0.1% bovine serum albumin (BSA; Sigma, St. Louis, U.S.A.). The sections were then incubated overnight with the primary antibodies in moist chambers at room temperature. The dilution of the primary antibodies was: NF-90, 1:10.000 (ascites); MAP 1A, 1:10.000 (ascites); MAP 5, 1:10.000 (ascites); MAP 2, 1:100 (culture supernatant); 3B9, 1:1000 (culture supernatant); V9, 1:1 (culture supernatant); and anti-GFAP, 1:500. The antibodies were diluted in PBS (pH 7.4) with 0.1% BSA and 1% normal goat serum (NGS; CLB, Amsterdam, the Netherlands), except V9 and anti-GFAP, which were both diluted in plain PBS (pH 7.4). Following the overnight incubation, the sections were rinsed three times in PBS (pH 7.4). Subsequently, the sections were incubated with the peroxidase-conjugated rabbit-anti-mouse antibody (RAM/PO; DAKO, Glostrup, Denmark) for 2 h under the same conditions as described for the primary antibody. In case of the anti-GFAP staining, GAR/PO antibodies were used (Nordic, Tilburg, the Netherlands). Both RAM/PO (1:500) and GAR/PO (1:300) were diluted in PBS (pH 7.4).

The sections were rinsed thoroughly in 50 mM TRIS-HCl buffer (pH 7.4) and developed in 80 mM 3,3'-diaminobenzidine-4-HCl (DAB; SERVA, Heidelberg, Germany) and 3 mM H_2O_2 in 50 mM TRIS-HCl buffer (pH 7.4). All sections were counterstained with hematoxylin for 3 s and rinsed with tap water for 10 min. Thereafter, the sections were dehydrated through upgraded alcohols to xylene and finally coverglassed with Entellan®.

Adjacent sections were stained with cresyl violet. Nonspecific immunohistochemical reactions and endogenous peroxidase activity were determined by incubation with preimmune normal mouse serum as primary antibody (in case of the anti-GFAP staining, preimmune normal rabbit serum was used). Each separate step of the immunocytochemical procedure was routinely checked for nonspecific reactions.

3.5 Enzyme Histochemistry

Nonfixed cryostat and fixed gelatin series were stained for the enzyme AChE. For both series, the same procedure was used for the detection of AChE, i.e., the Karnovsky and Roots method (1964) with acetylthiocholine (AThCh) as the substrate. In short, cryostat sections were incubated for 3 h at room temperature in a mixture of 5 mM sodium citrate (Merck, Darmstadt, Germany), 3.5 mM copper sulfate (Merck), 0.5 mM potassium ferricyanide (Brocacef, Maarssen, the Netherlands), and 2 mM AThChiodide (Merck) dissolved in a 50 mM TRIS-maleic buffer (pH 6.0). Only the application of the appropriate enzyme inhibitors to the incubation medium will guarantee reliable results, since AThCh can also be converted by other enzymes present in central nervous tissue. In this study, the enzyme inhibitors eserine (for the nonspecific esterases) and tetraisopropylphosphoramide (iso-OMPA; for the pseudocholinesterase, pseudo-ChE) were added to the standard incubation medium. Both inhibitors were first tested in concentrations ranging from $10^{-9} M$ to $10^{-2} M$. Eserine and iso-OMPA were found to produce reliable results in a $10^{-5} M$ concentration (see also Marani et al. 1977; Marani 1981).

After incubation, the sections were rinsed three times in distilled water, followed by a 5-min fixation in 4% paraformaldehyde in 0.1 M phosphate buffer (pH 7.3). Afterwards, the nonfixed cryostat sections were thoroughly rinsed with distilled water and coverslipped with gelatin–glycerine (1:3). The free-floating fixed gelatin sections were first mounted on chrome alum–gelatin-coated slides. After being air-dried overnight, these sections were dehydrated through upgraded alcohols to xylene and finally coverglassed with Entellan®. It should be emphasized that fixed gelatin sections of the younger embryonal rats (E9–E16) were incubated for 6 and 22 h, in addition to the standard 3 h (see Sect. 4.1). In each series, adjacent sections were stained with cresyl violet.

3.6 Neuronal Tracing: Intrauterine and Postnatal Applications

Wistar Albino Glaxo rats in estrus were mated between 10 and 11 A.M. (van der Zwet et al. 1986). The end of this period was taken as the start of E0 (Paxinos et al. 1990). In our colony, pregnancy lasts 22 days, i.e., the pups are born at the end of E21 or early in E22. At the end of required duration of pregnancy, the females (body weight 250–300 g) were anesthetized by intraperitoneal injection of 0.4 ml Valium® (5 mg/ml diazepam; Hoffmann-La Roche, Basle), and 0.15 ml Hypnorm® (0.2 mg/ml fentanyl; Ceva, Paris). A subcutaneous injection of 0.1 ml atropine sulphate (500 µg/ml) was given to diminish mucous secretion into the tracheobronchial tree.

A midline laparotomy was performed to expose the uterine horns. After one of the uterine horns had been exteriorized, the position of a fetus was determined by cold-light transillumination. A single injection of HRP or WGA–HRP (Sigma grade VI, Amsterdam) was made directly through the translucent uterine wall, aimed at the spinal cord of the fetus and using the dorsal spinal artery as a landmark. All injections were made through glass micropipettes connected to a pressure ejection system (Rogers 1985). Ejection pressure and

Table 2. Developmental age at injection, injected tracer, and developmental age at perfusion of the rat fetuses and pups described

Perfusion		Injection		Perfusion		Injection	
Age	C number	Age	Tracer	Age	C number	Age	Tracer
E17	C3523	E16	5% WGA-HRP	E20	C3657.3	E19	5% WGA-HRP
	C3531	E16	5% WGA-HRP		C4078.5	E17	40% HRP
	C4107.3	E16	40% HRP		C4255.2	E19	40% HRP
	C4258.2	E16	40% HRP		C4263.3	E17	40% HRP
	C4258.3	E16	40% HRP		C4263.4	E17	40% HRP
	C4258.6	E16	40% HRP		C4273.5	E17	40% HRP
E18	C3315	E17	5% WGA-HRP	E21	C3180	E20	5% WGA-HRP
	C3316	E17	5% WGA-HRP		C3934.2	E19	5% WGA-HRP
	C4098.5	E17	40% HRP		C3934.3	E19	5% WGA-HRP
	C4109.2	E17	40% HRP		C4187.3	E19	40% HRP
	C4157.1	E17	40% HRP		C4187.5	E19	40% HRP
	C4162.1	E17	40% HRP		C4211.2	E19	40% HRP
	C4162.3	E17	40% HRP	P2	C4243.1	P1	5% WGA-HRP
	C4169.1	E17	40% HRP		C4243.2	P1	5% WGA-HRP
E19	C3252	E18	5% WGA-HRP		C4243.4	P1	5% WGA-HRP
	C3253	E18	5% WGA-HRP	P3	C4245.1	P2	5% WGA-HRP
	C3945.1	E18	40% HRP		C4245.2	P2	5% WGA-HRP
	C3945.6	E18	40% HRP		C4245.3	P2	5% WGA-HRP
	C4011.3	E17	40% HRP		C4245.4	P2	5% WGA-HRP
	C4324.4	E18	5% WGA-HRP	P4	C3208	P3	40% HRP
	C4324.6	E18	5% WGA-HRP		C4247.1	P3	5% WGA-HRP
E20	C3657.1	E19	5% WGA-HRP		C4247.2	P3	5% WGA-HRP
	C3657.2	E19	5% WGA-HRP		C4247.3	P3	5% WGA-HRP

WGA, wheat germ agglutinin; HRP, horseradish peroxidase; E17–21, embryonal days 17–21; P2–4, postnatal days 2–4.

Table 3. Age at the time of perfusion, C number, injection sites, and labeling scores of all experimental animals described.

Injection site columns: Spinal cord = Lumbosacral, Thoracic, Cervical; plus MO. Labeling columns: VN, RN, TLD, LC (v, d), NR (c, vl, dm, pc). A ▬ marks the rostrocaudal extent of the injection bar.

Perfusion Age	C number	Lumbosacral	Thoracic	Cervical	MO	VN	RN	TLD	LC v	LC d	NR c	NR vl	NR dm	NR pc
E17	C4258.6			▬	▬	×	×	×	×	×	+	+	±	±
	C3523			▬	▬	+	+	+	+	+	+	+		
	C4107.3			▬		+	+	+	+					
	C3531			▬		+	+	+						
	C4258.3		▬	▬		+	+	+						
	C4258.2	▬				+								
E18	C3315			▬	▬	+	+	+	+	+	+	+	±	±
	C3316			▬		+	+	+	+	+	+	+		
	C4162.3			▬		+	+	+	+	+				
	C4109.2		▬	▬		+	+	+	+		+	±		
	C4098.5	▬	▬			+	+	×	+		+			
	C4157.1		▬			+	+	+			±			
	C4169.1		▬			+	+	+						
	C4162.1	▬	▬			+	+							
E19	C3945.1			▬	▬	+	+	+	+		+	+	+	+
	C3252		▬	▬	▬	+	+	+	+		+	+	±	±
	C3253			▬		+	+	+	+		+	+	±	±
	C4011.3			▬		+	+	+	+					
	C4324.4		▬	▬		+	+	+	+		+	+	±	
	C4324.6	▬	▬			+	+	+	+		+	+		
	C3945.6	▬				+	+	+						
E20	C3657.1		▬	▬	▬	+	+	+	+		+	+	+	+
	C3657.2		▬	▬		+	+	+	+		+	+		
	C4255.2		▬	▬		+	+	+	×		+	+		
	C4078.5			▬		+	+	+			+	+		
	C4273.5		▬			+	+	+	+		+	±		
	C4263.3		▬			+	+	+	+		±			
	C4263.4	▬	▬			+	+	+	+					
	C3657.3	▬	▬			+	+	+	+					
E21	C3934.3			▬	▬	+	+	+	+		+	+	+	+
	C3180			▬		+	+	+	+		+	+	+	+
	C3934.2	▬	▬			+	+	+	+		+	+		
	C4211.2	▬	▬			+	+	+	+		+	+		
	C4187.5	▬				+	+	+	+		+	+		
	C4187.3	▬				+	+	±						
P2	C4243.1			▬		+	+	+	+		+	+	+	+
	C4243.2	▬				+	+	+	+		+	+		±
	C4243.4	▬				+	+	+	±		+	+		±
P3	C4245.4		▬	▬		+	+	+	+		+	+	±	+
	C4245.2	▬	▬			+	+	+	+		+	+	±	+
	C4245.3	▬	▬			+	+	+	+		+	+		+
	C4245.1	▬				+	+	+	+		+	+		+
P4	C3208			▬	▬	+	+	+	+		+	+	+	+
	C4247.3			▬	▬	+	+	+	+		+	+	+	+
	C4247.1	▬				+	+	+	+		+	+		±
	C4247.2	▬	▬			+	+	+	+		+	+		±

+, Labeled neurons present throughout the subgroup; ±, labeled neurons only present in a part of the subgroup; ×, indeterminable due to technical causes (folded section, etc.). MO, medulla oblongata; VN, vestibular nuclei; RN, raphe nuclei; TLD, nucleus tegmentatis laterodorsalis; LC, locus coeruleus; NR, nucleus ruber; v, ventral; d, dorsal; c, caudal; E17–21, embryonal days 17–21; P2–4, postnatal days 2–4. vl, ventrolateral; dm, dorsomedial; pc, parvocellular.

length of ejection phase were gauged before injection to yield an injected volume of approximately $0.1\,\mu l$ (Rogers 1985). However, because penetration and injection were performed manually, the actually injected volume was variable. WGA–HRP was injected as a 5% solution, and HRP as a 40% solution; both tracers were dissolved in $0.05\,M$ TRIS-maleic acid buffer (pH 7.6). Up to six fetuses were injected per pregnant female, and up to three per uterine horn.

After a suitable survival time, the laparotomy was repeated, and the injected fetuses were removed from the uterus one by one. Each fetus was perfused transcardially with 5–10 ml lukewarm normal saline, followed by 5–20 ml citrate buffer $(0.1\,M;\,\mathrm{pH}\,7.1)$ containing 1.25% glutaraldehyde and 1% paraformaldehyde. After perfusion, the central nervous system was removed in toto and embedded in 13% gelatin (Difco, Detroit). Transverse 40-μm sections were cut on a Jung freezing microtome. Every second or fourth section was processed according to the tetramethylbenzidine (TMB) method (Mesulam 1982). The sections were mounted on chrome alum-subbed slides, counterstained with neutral red, and cover-slipped with Permount® mounting medium (Fisher Scientific, New Jersey). The sections were examined under bright and dark field illumination. Cells were considered as labeled when they met the criteria described by Nauta et al. (1974). Unsuccessful experiments (no injection site present) served as controls for endogenous peroxidase activity.

In addition to the prenatal experiments, rat pups with ages ranging from P0 (day of birth) to P4 were operated upon. These rat pups were anesthetized by intraperitoneal injection of $6\,\mu l$ Valium and $2\,\mu l$ Hypnorm. A laminectomy was performed at the level of the cervical or lumbar enlargement, and WGA–HRP was injected into the spinal cord. Upon recovery from the anesthetic, the pups were returned to the litter. After a survival period of 1 day, the pups were reanesthetized and processed in the same way as the fetuses (see above).

Series in which no injection site was present in the spinal cord or in which leakage had occurred into the central canal or subarachnoid space were discarded. The presence of retrogradely labeled neurons in both the medullary raphe nuclei (pallidus, obscurus, and magnus) and the vestibular nuclei served as a control for false negative results caused by insufficient transport of the tracer due to imponderabilia. Axons of serotoninergic neurons in the medullary raphe nuclei were reported to be present at all levels of the spinal cord at E17 and in the dorsal horn of all levels at E19 (Rajaofetra et al. 1989); vestibular axons were reported to be present in the lower spinal cord at E17 (?) (Auclair et al. 1991). Series were included only if retrograde labeling was present in the medullary raphe nuclei and in the vestibular nuclei.

Even after these precautions, insufficient transport time might still be a source of false negative results, especially so in the case of nuclei located rostrally to the (retrogradely labeled) medullary raphe nuclei. For this reason, series which did not contain retrogradely labeled neurons in the pertinent nucleus were only included if labeled neurons were present at more rostral levels of the central nervous system.

The injection sites of these 46 series were reconstructed from serial camera lucida drawings of the spinal cord sections. All sections in which evenly spread, fine-grained TMB deposit was present were included in the injection site. The injection sites were projected onto diagrams of the spinal cord. In most cases, the distribution of TMB deposit through the spinal cord was bilaterally symmetrical. The experiments described in the present paper are listed in Tables 2 and 3. Experiments with injection sites located below the rostralmost "negative" injection were omitted from these tables.

4 Results

4.1 General Comments on the Techniques

The control experiments demonstrated a high specificity of the antibodies in the various stages of the immunohistochemical procedure. Sections of spinal cords from rats of different ages incubated with preimmune normal mouse serum or preimmune normal rabbit serum were found to be negative. Endogenous peroxidase could only be detected in erythrocytes in those animals that were not adequately, or not at all, perfused. Additionally, endogenous catalase activity could often be found in the pial blood vessels.

The present results demonstrated regional differences in the intensity of NF-90 immunostaining in the spinal cord funiculi. This phenomenon has been described earlier (Oudega et al. 1990b). In a recent study, the immunocytochemical staining of neurofilament with the antibody NF-90 in order to detect nerve fibers was compared with the classical Bodian fiber staining (Oudega et al. 1990b). The latter is known to identify the three neurofilament subunits (Gambetti et al. 1981; Autillio-Gambetti et al. 1986; Oudega et al. 1990b). Although quantification has not been performed, it seemed that in the embryonal spinal cord more intense NF-90 staining represented a higher number of fibers. More intense NF-90 staining could also be detected in the postnatal rat spinal cord and in most cases appeared to reflect the presence of large-caliber fibers (Oudega et al. 1990b).

The MAP 1A antibody (clone BW6) used in this study recognizes the MAP 1A subspecies (Garner et al. 1990). The present results demonstrated transient expression of MAP 1A in the different layers of the developing rat spinal cord. More surprisingly, however, was the early appearance of MAP 1A, since this particular MAP is considered to be a member of the group of late MAP. Nevertheless, as early as E12, MAP 1A could be detected in the matrix and mantle layer of the spinal cord (see Oudega et al. 1990b).

MAb 3B9 is directed against the carbohydrate FAL, also known as SSEA-1, Le^x, X-hapten, and CD15 (Solter and Knowles 1979; Gooi et al. 1981; Fox et al. 1983; Feizi 1985). Antibodies recognizing CD15 show considerable heterogeneity in their specificity and biological properties (Mai and Reifenberger 1988; Bartsch and Mai 1991; Oudega et al. 1992; Plank and Mai 1992). This heterogeneity is most likely caused by the recognition of different epitopes on FAL, which can be created by binding of this trisaccharide to different carriers. Recently, it was shown that 3B9 and anti-Lue-M1 (see Mai and Reifenberger 1988; Bartsch and Mai 1991; Plank and Mai 1992) actually recognize the same epitope, but in different spatial configurations (Marani et al. 1992a,b). In this study, the term SSEA-1 refers to the FAL (or CD15) epitope, as recognized by the antibody 3B9.

SSEA-1 was identified in the ectoderm layer of E9 rat embryos, which was the youngest age studied (see also Oudega et al. 1992). On this day, as well as the neural groove, other non-neural parts of the ectoderm layer also appeared to be positive for SSEA-1. During the following days, 3B9 immunoreactivity gradually diminished from the non-neural ectoderm (Oudega et al. 1992). In this study, the presence of the SSEA-1 antigen in the ectoderm will not be emphasized.

Especially in the spinal cord of the younger rat embryos, a marked difference was demonstrated between the expression patterns of AChE in non-fixed cryostat sections and in fixed gelatin sections (Oudega and Marani 1990). Such findings were also described for other parts of the brain and underline the necessity of using both techniques in developmental, anatomical studies on the localization of AChE in the central nervous system (Marani and Voogd 1977; Oudega and Marani 1990). Although anatomical features appear to be better preserved after an aldehyde fixation, the nonfixed sections proved to be more accurate in showing the early distribution of AChE (Oudega and Marani 1990). For the description of the present results, the fixed gelatin series were used because of their superior tissue preservation. Nevertheless, additional findings in the nonfixed cryostat series will be emphasized. Sections of the spinal cord of especially the younger embryos (E10–E16) were incubated for a period of 6 or even 22 h in addition to the routinely used 3-h period. This was necessary since the 3-h incubation period appeared to be inconsistent in revealing the accurate AChE distribution patterns in the younger embryos. Such an inconsistency could be caused by low amounts of the enzyme in the spinal cord during early embryonal development. A prolonged incubation period could cause a better penetration of the substrate AThCh into the tissue and indeed resulted in a more accurate presentation of the AChE distribution patterns (Oudega and Marani 1990).

The overall presence of pseudo-ChE activity in the spinal cord and especially in a subpopulation of AChE-positive motor neurons has been described earlier (Koelle 1954; Oudega and Marani 1990). Additionally, nonspecific esterases were found to be present in the cell lining of blood vessels (see Marani 1981), in a subpopulation of the motor neurons and in the cells of the intermediolateral column of the postnatal spinal cord (Söderholm 1965; Oudega and Marani 1990). The present study confirmed all these findings and stresses the necessity of enriching the incubation medium with enzyme inhibitors such as iso-OMPA and eserine in order to inhibit pseudo-ChE and nonspecific esterases, respectively.

Within the spinal cord, MAb V9 demonstrated the presence of vimentin in the cell lining of the blood vessels and in the meningeal membranes. These findings were reported earlier (Joosten and Gribnau 1989; Oudega and Marani 1991). In this study, the blood vessels will not be discussed. The presence of vimentin in the meningeal membranes, however, will be described and discussed because of their apparent participation in the glial system of the rat spinal cord.

The PAb anti-GFAP was found to recognize its antigen in the longitudinal fibers of the spinal cord funiculi. Most likely, their antigenic determinant is also present on the neurofilaments which are incorporated in the axonal cytoskeleton. Cross-reactivity of this particular anti-GFAP antiserum (DAKO no. Z 334) with neurofilament proteins has been suggested before (Hansen et al. 1989; Oudega and Marani 1991). In this study, the glial fibers in the rat spinal cord funiculi demonstrated a unique staining pattern and could, therefore, easily be distinguished from the neuronal processes.

During the first 3 postnatal weeks, a transition from vimentin expression to GFAP expression was found to occur in the developing rat spinal cord. This transition period has been noticed before (Joosten and Gribnau 1989; Oudega and Marani 1991). During this period, both proteins appeared to be simultaneously expressed in the same glial cell (Schnitzer et al. 1981; Pixley and De Vellis 1984; Joosten and Gribnau 1989). The vimentin–GFAP transition period, as found in this study, coincides with the earlier described time span. However, from the present results it could not be determined whether the proteins were simultaneously present in the same glial cell.

In general, the time of appearance and the distribution patterns of the studied antigens and enzymes in the developing rat spinal cord appeared to demonstrate a rostral-to-caudal gradient (see also Oudega and Marani 1990, 1991; Oudega et al. 1990a,b, 1992). Previously, such a gradient was also found to be present in the development of the cytoarchitecture of the rat spinal cord (Altman and Bayer 1984). Depending on the examined region or feature within the spinal cord, the ventral-to-dorsal gradient was found to develop during different time periods.

Although all the studied markers were examined in serial sections of the cervical, thoracic, and lumbar spinal cord, this report will especially describe their distribution patterns at the thoracic level. Differences in the appearance or localization of the markers in other levels of the cord due to technical or developmental aspects will be described and discussed. In addition, special anatomical features which are restricted to certain spinal cord levels will be emphasized.

To describe the localization of the different markers, an arbitrary division of the developing rat spinal cord was used (see Fig. 2). This division is purely anatomical and mainly based on the thymidine autoradiographic experiments of Altman and Bayer (1984). One has to bear in mind that such a subdivision never reflects the proper functional relationships of the different spinal cord regions. For reasons of clarity, different anatomical regions or special anatomical features of the spinal cord are described together. The spinal cord matrix layer (or ependymal layer) was divided into the basal plate/floor plate and the alar plate/roof plate. The mantle layer of the spinal cord (or gray matter) was divided into the ventral horn/ventral commissure/ventral raphe, the intermediate gray, and the dorsal horn/dorsal raphe. The development of the spinal cord lateral horn, which contains the autonomic intermediolateral nucleus, will be described as part of the intermediate gray. Finally, the marginal layer (or white matter) was divided into the ventral funiculus, the lateral funiculus, and the dorsal funiculus/dorsolateral fasciculus. It should be noted that during early embryogenesis of the spinal cord, the dorsal funiculus develops from the dorsal root entrance zone and the dorsal root bifurcation zone (see Altman and Bayer 1984). Later on in development, the medial fasciculus gracilis and the lateral fasciculus cuneatus can be distinguished from each other in the dorsal funiculus. Additionally, the rat dorsal funiculus includes the corticospinal tract, which is situated in the most ventral part of the dorsal columns. All these regions will be described and discussed as part of the dorsal funiculus.

For each region or combination of regions, the appearance and expression patterns of the studied markers will be described in a particular sequence. Where appropriate, the results will be preceded by a brief description of the development of the cytoarchitecture of the particular region, taken in particular from the work of Altman and Bayer (1984) and Nornes and Das (1974).

First, those markers associated with the neuronal cytoskeleton (neurofilaments and MAP) will be described. Next, the ontogeny of the so-called functional markers (SSEA-1 and AChE) will be reported, followed by the description of the appearance and expression patterns of the markers associated with the glial cytoskeleton (vimentin and GFAP). Finally, separately from the immunocytochemical findings, the results of the intrauterine and postnatal neuronal tracer experiments will be reported.

4.2 Matrix Layer

4.2.1 Basal Plate/Floor Plate

4.2.1.1 Cytoarchitecture

Between E10 and E14, after the closure of the neural tube, the cells of the ventral part of the matrix layer proliferate. Especially during the period E11–E12, an explosive proliferation can be observed in the so-called basal plate. During the mitotic cycles, the cells remain in close contact with the internal and external membranes of the matrix layer (known as the membrana limitans interna and externa, respectively) by way of cytoplasmatic extensions. The nuclei of the proliferating cells migrate along these extensions, which gives the matrix layer its stratified appearance. After a certain number of mitotic cycles, the cells release their contacts and migrate towards the periphery of the basal plate. In the ventral matrix layer, apart from neuroblasts, glioblasts are also generated, which later develop into microglial cells, astrocytes, or oligodendrocytes (see Bondar 1977). These three cell types form the future glial system of the rat spinal cord.

In cresyl violet-stained sections of the spinal cord of the younger embryos, the matrix layer appeared to be mainly occupied by darkly stained, closely packed cells. Later on in development, the remaining cells in the matrix layer were found to be much less intensely stained for the Nissl substance.

4.2.1.2 Neurofilaments

Throughout the embryonic and the postnatal period of spinal cord development, NF immunoreactivity was found to be absent from the basal plate and the floor plate of the matrix layer.

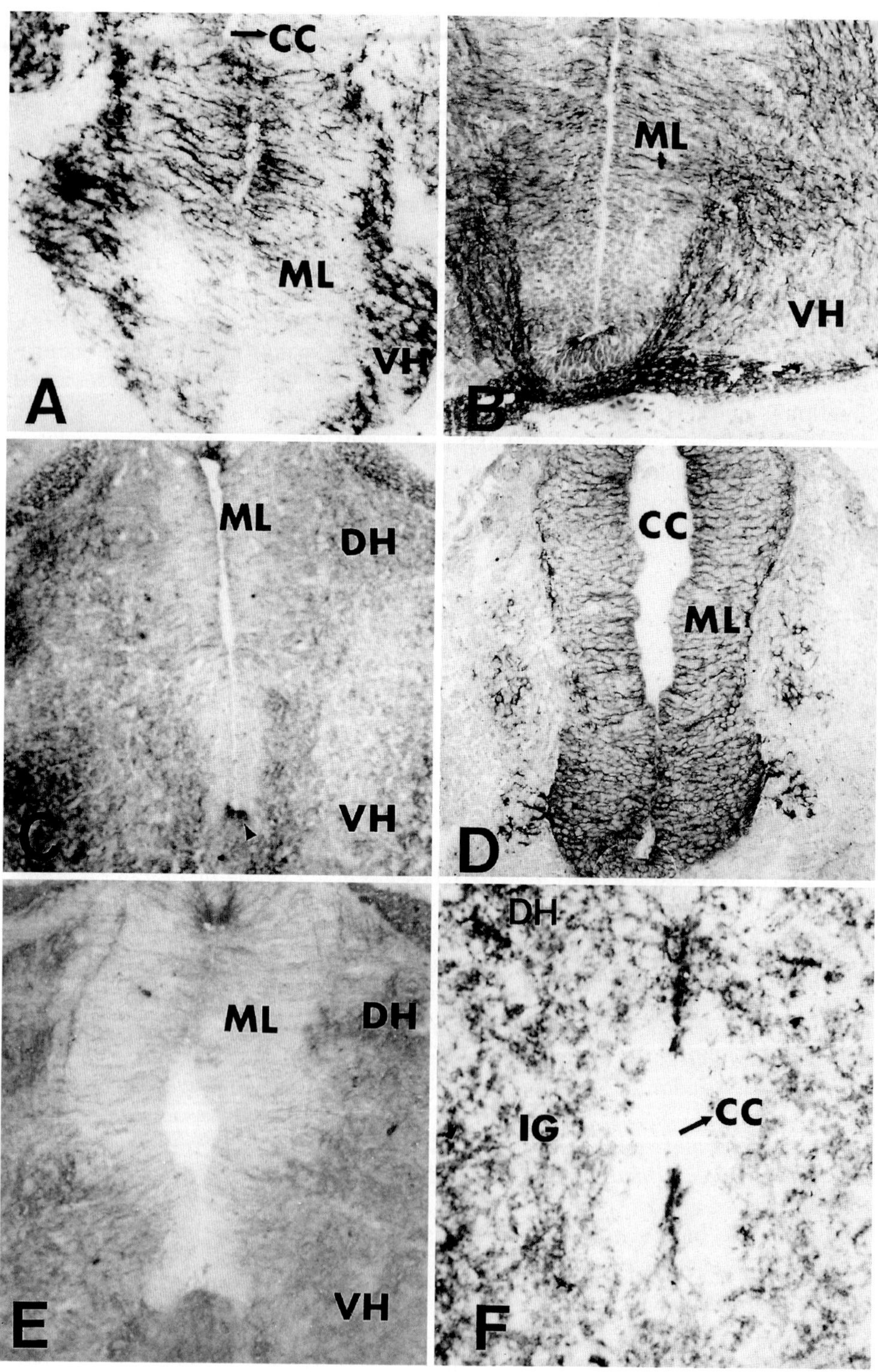

CC
ML
VH
A
ML
VH
B
ML
DH
VH
C
CC
ML
D
ML
DH
VH
E
DH
IG
CC
F

4.2.1.3 Microtubule-Associated Proteins

At E12, a radial pattern of moderately stained MAP 1A-positive processes were detected throughout the spinal cord basal plate (Fig. 4A). Two days later, at E14, almost all the MAP 1A-positive process disappeared from the ventral matrix layer (Fig. 4B). At E17, the basal plate appeared to be completely devoid of MAP 1A immunoreactivity (Fig. 4C).

Between E13 and E15, MAP 1A staining was detected in the floor plate. Around E15, a patch-like MAP 1A staining was found on each side of the imaginative midline (Fig. 4C). Around E17, the antigen could hardly be detected, and around E20 all the MAP 1A immunoreactivity had vanished from the floor plate.

During the development of the rat spinal cord, MAP 2 immunoreactivity could not be demonstrated in the basal plate or in the floor plate. In the adult rat spinal cord, both regions also appeared to be devoid of MAP 2.

At E12, an abundant MAP 5 radial staining pattern was found in the basal plate of the rat spinal cord (Fig. 4D). During the following days, the reactivity for the MAP 5 antigen rapidly diminished, and at E16 hardly any positive process could be identified in the basal plate (Fig. 4E).

At E12, the floor plate appeared to be intensely stained for the MAP 5 antigen (Fig. 4D). Until P2, a small MAP 5-positive area was detected in the floor plate (Fig. 4F). Thereafter, the intensity of the staining rapidly decreased, and after P4 MAP 5 immunoreactivity was found to be absent from the spinal cord floor plate.

4.2.1.4 Stage-Specific Embryonic Antigen-1

In E11-aged rat embryos, a high 3B9 immunoreactivity was detected in the vicinity of the ventricle of the basal plate of the spinal cord (Fig. 5A). In fact, 1 day earlier this specific localization of SSEA-1 could already be detected in the then closing neural tube. At E12, the staining intensity in the ventral part of the ventricular region decreased, and after E13 the antigen could not be detected anymore in this region. During the embryonal and postnatal development of the rat spinal cord, the floor plate appeared to be devoid of SSEA-1 immunoreactivity (Fig. 5A).

4.2.1.5 Acetylcholinesterase

At E11, AChE activity was found to be present in the basal plate of the rat spinal cord (Fig. 5B). During the following days, AChE activity increased in the ventral matrix layer. At E16, the stained area was found to be sharply

Fig. 4A–F. Immunocytochemical localization of microtubule-associated proteins in the developing ventral matrix layer (*ML*) and floor plate of the rat spinal cord. Microtubule-associated protein 1A at embryonal day 12 (E12; **A**), E14 (**B**), and E15 (**C**). Microtubule-associated protein 5 at E12 (**D**), E16 (**E**), and E20 (**F**). *CC*, central canal; *VH*, ventral horn; *DH*, dorsal horn; *IG*, intermediate gray

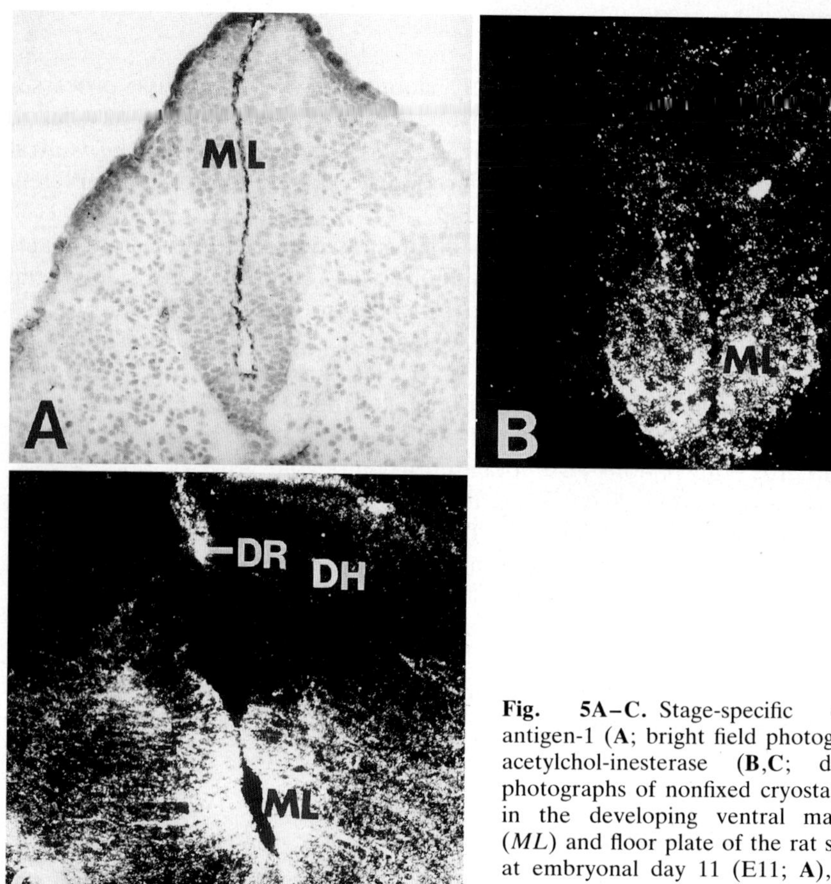

Fig. 5A–C. Stage-specific embryonic antigen-1 (**A**; bright field photograph) and acetylchol-inesterase (**B,C**; dark field photographs of nonfixed cryostat sections) in the developing ventral matrix layer (*ML*) and floor plate of the rat spinal cord at embryonal day 11 (E11; **A**), E11 (**B**), and E16 (**C**). *DR*, dorsal raphe; *DH*, dorsal horn

delineated by the sulcus limitans (Fig. 5C). AChE activity could not be detected transgressing this landmark. It should be stated that the sulcus limitans could not be detected at all spinal cord levels. After E16, the AChE staining intensity rapidly decreased, and 2 days later the enzyme was completely abolished from the basal plate. This particular AChE expression pattern could only be detected in the nonfixed cryostat sections. In the fixed gelatin sections of the developing rat spinal cord, the enzyme could not be detected in the basal plate.

At E11, the floor plate of the spinal cord matrix layer appeared to be stained for AChE (Fig. 5B). Three days later, at E14, the floor plate was devoid of the enzyme. This temporary presence of AChE in the spinal cord floor plate could only be detected in the non-fixed cryostat sections.

4.2.1.6 Vimentin

In E11-aged rat embryos, a scattered vimentin staining was found to be present in the basal plate. At the light microscopic level, the exact localization of this

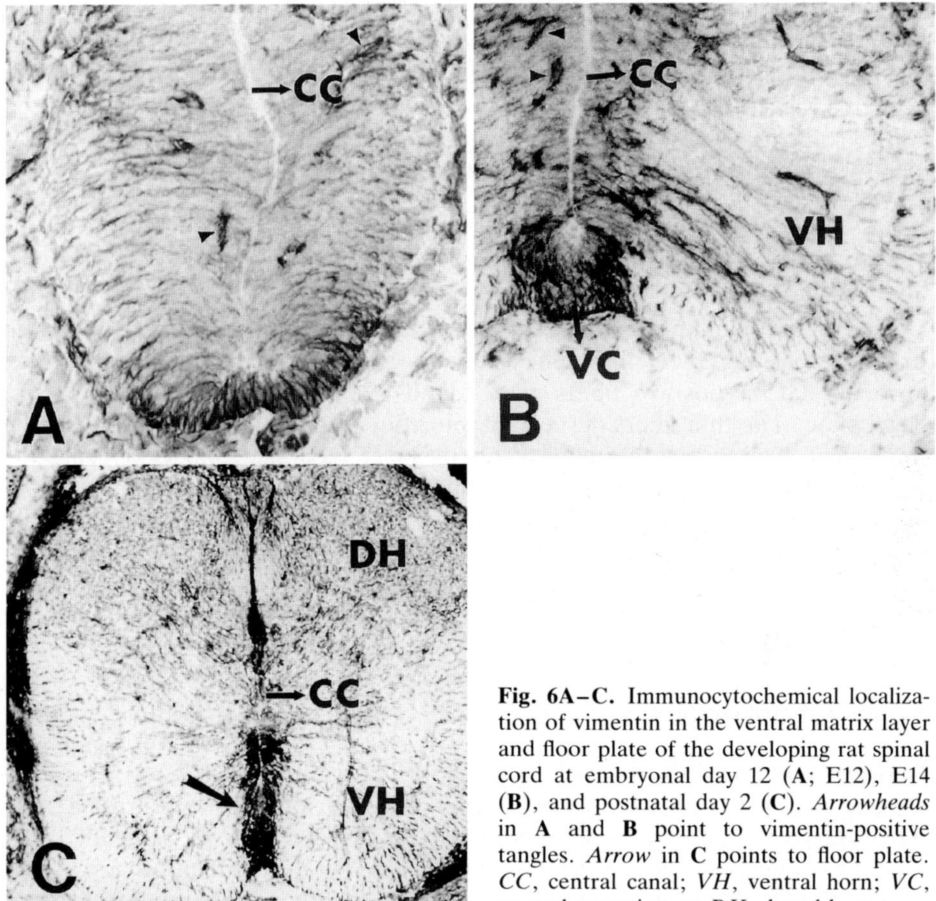

Fig. 6A–C. Immunocytochemical localization of vimentin in the ventral matrix layer and floor plate of the developing rat spinal cord at embryonal day 12 (**A**; E12), E14 (**B**), and postnatal day 2 (**C**). *Arrowheads* in **A** and **B** point to vimentin-positive tangles. *Arrow* in **C** points to floor plate. *CC*, central canal; *VH*, ventral horn; *VC*, ventral commissure; *DH*, dorsal horn

faint vimentin immunoreactivity was difficult to determine. However, 1 day later, the now more abundant vimentin positivity was accurately localized in the membrana limitans externa (the external border of the neural tube). Abundant vimentin staining was also found throughout the basal plate of the rat spinal cord. During the following days of development, the vimentin immunoreactivity developed into a more regular pattern (Fig. 6A). Although some scattered short fibers were still present, most of the vimentin staining of the basal plate appeared to be distributed in an organized radial pattern. Most of the vimentin-positive fibers of the basal plate coursed after a sharp curve in the direction of the ventral commissure (Fig. 6B; see Sect. 4.3.1.6). Others coursed in the direction of the ventral mantle layer. The described staining pattern remained present after birth (Fig. 6C). The staining intensity, however, rapidly decreased and at P4 the basal plate was found to be devoid of vimentin immunoreactivity. Surprisingly, after P10 and especially in the mature spinal cord, short, twisting, vimentin-positive fibers were found to emerge from the basal plate in all directions into the mantle layer.

33

At E12, short, vimentin-positive protrusions appeared to penetrate the spinal cord floor plate (Fig. 6A). These positive fibers seemed to originate in the membrana limitans externa. During the following days, the floor plate developed a clear radial vimentin staining. This staining pattern was detected until birth; thereafter, the intensity rapidly decreased and around P4 the floor plate was found to be devoid of vimentin immunoreactivity.

4.2.1.7 Glial Fibrillary Acidic Protein

During embryonic development, the basal plate of the rat spinal cord appeared to be devoid of GFAP immunoreactivity. During the first postnatal week, however, GFAP-positive fibers were found to be present in the spinal cord basal plate. The thin fibers curved sharply after leaving the layer and formed a small ventral raphe in the ventral mantle layer (see Sect. 4.3.1). During the embryonal as well as the postnatal period of spinal cord development, GFAP immunoreactivity was found to be absent from the floor plate of the rat spinal cord matrix layer.

4.2.2 Alar Plate/Roof Plate

4.2.2.1 Cytoarchitecture

Between E13 and E16, the cells of the dorsal part of the matrix layer proliferate and establish a pool of future interneurons. During the period E14–E15 in particular, an explosive proliferation can be observed in the so-called alar plate. As already described for the basal plate, during their mitotic cycles the cells of the alar plate also remain in close contact with the internal and external membranes of the matrix layer by way of cytoplasmatic extensions. Their nuclei migrate along these extensions, which gives the alar plate its stratified appearance. After a certain number of mitotic cycles, the cells release their contacts and migrate towards the periphery of the alar plate. As well as the future neurons, glioblasts are also generated within the alar plate. These cells develop into microglial cells, astrocytes, or oligodendrocytes, which later form the glial system of the rat spinal cord (Bondar 1977). In younger embryos, the cells of the dorsal matrix layer can be identified as darkly stained, closely packed cells in cresyl violet sections. Later on in development, the remaining cells were found to stain much less intensely for the Nissl substance. It should be mentioned that in the following description, the intermediate region of the matrix layer actually represents the area near the sulcus limitans.

Fig. 7A–F. Expression of microtubule-associated proteins 1A, 2, and 5 in the developing dorsal matrix layer (*ML*) and roof plate of the rat spinal cord. Microtubule-associated protein 1A at embryonal day 13 (**A**; E13) and E17 (**B**). **C** Microtubule-associated protein 2 at E16. Microtubule-associated protein 5 at E13 (**D**), E16 (**E**), and E20 (**F**). *CC*, central canal; *DH*, dorsal horn; *IG*, intermediate gray

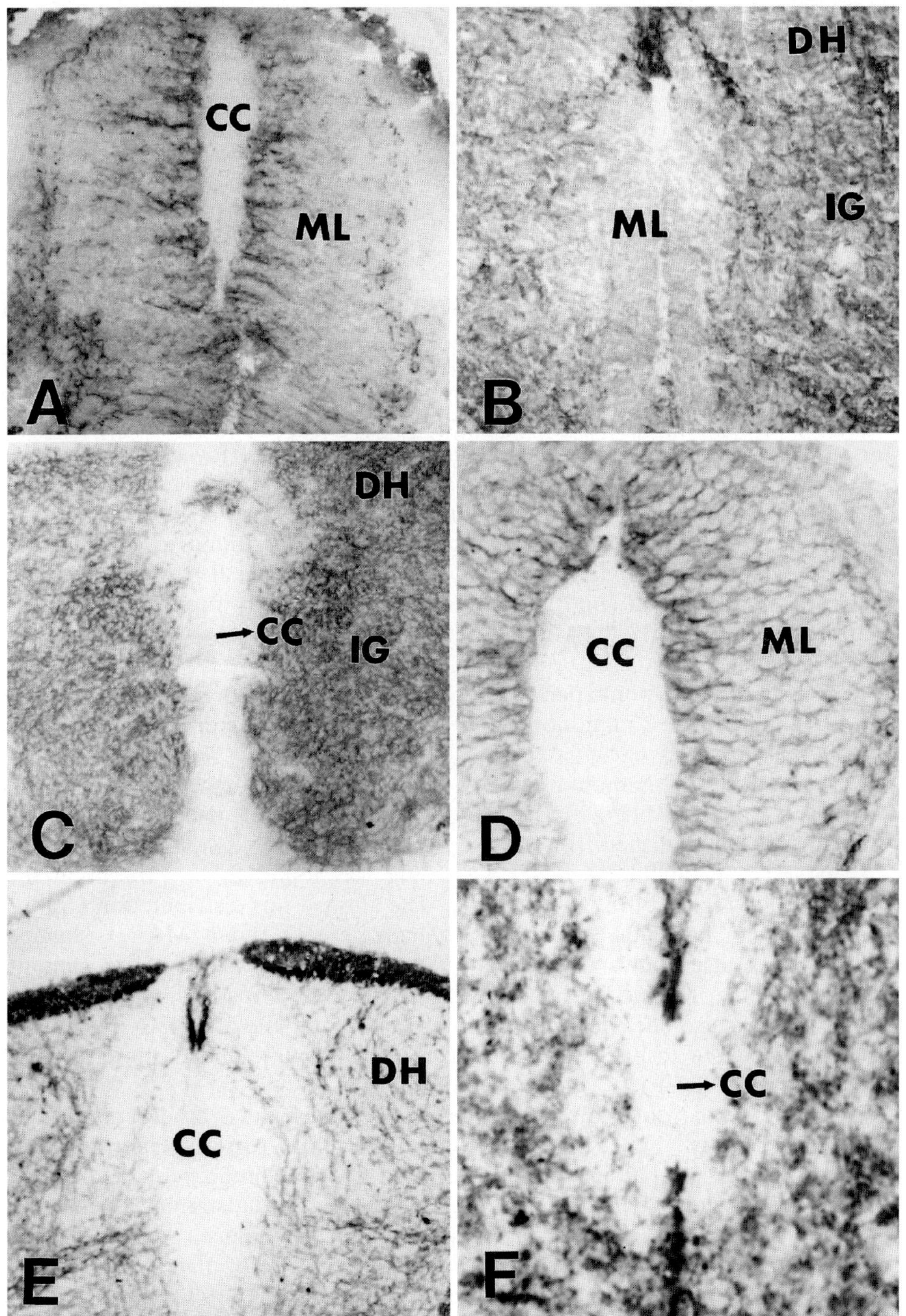

CC
ML
A
DH
ML
IG
B
DH
CC
IG
C
CC
ML
D
DH
CC
E
CC
F

4.2.2.2 Neurofilaments

Throughout the embryonal and postnatal development of the rat spinal cord, NF immunoreactivity appeared to be absent from the alar plate and the roof plate of the matrix layer.

4.2.2.3 Microtubule-Associated Proteins

At E13, moderate MAP 1A staining was found in the alar plate of the spinal cord (Fig. 7A). The antigen appeared to be distributed in a distinctly radial pattern. At this age, the intermediate region of the matrix layer also stained positively for MAP 1A. One day later, at E14, the moderately stained radial pattern persisted in the alar plate, but was found to be especially prominent in the intermediate area. During the following days, the pattern remained essentially similar, but the staining intensity diminished towards a faint appearance. At E17, MAP 1A appeared to be completely abolished from the alar plate as well as the intermediate region of the matrix layer (Fig. 7B).

At E13, the roof plate of the rat spinal cord was found to demonstrate MAP 1A immunoreactivity (Fig. 7A). Around E17, a small MAP 1A-positive area was still detected (Fig. 7B). After E21, the activity rapidly diminished, and around P4 the MAP 1A antigen had completely vanished from the spinal cord roof plate.

During the embryonic period as well as the postnatal period of spinal cord development, MAP 2 immunoreactivity could not be detected in the alar plate of the matrix layer.

At E16, MAP 2 immunoreactivity was found to be present in the roof plate of the rat spinal cord (Fig. 7C). Two days later, however, the MAP 2 staining appeared to be abolished from this region.

At E12, an abundant MAP 5 staining was found to be present in the intermediate region of the matrix layer. The antigen was distributed in a radial pattern. One day later, at E13, an equally intense radial MAP 5 staining pattern was also found in the alar plate (Fig. 7D). At E14, the staining intensity in the intermediate part of the matrix layer already diminished. One day later, this also became apparent in the alar plate, and after E16 the MAP 5 antigen appeared to be absent both from the alar plate and the intermediate matrix layer (Fig. 7F).

At E13, the roof plate was found to demonstrate a moderate MAP 5 immunoreactivity (Fig. 7D). Three days later, this region demonstrated a patch-like staining pattern (Fig. 7E). An intense staining remained visible between E16 and P2 (Fig. 7F). Thereafter, however, the staining intensity of these patches rapidly diminished towards a faint appearance. Around P6, the antigen had completely vanished from the spinal cord roof plate.

4.2.2.4 Stage-Specific Embryonic Antigen-1

Between E11 and E13, SSEA-1 was found in the alar plate and in the intermediate region of the matrix layer. There was strong immunoreactivity in the

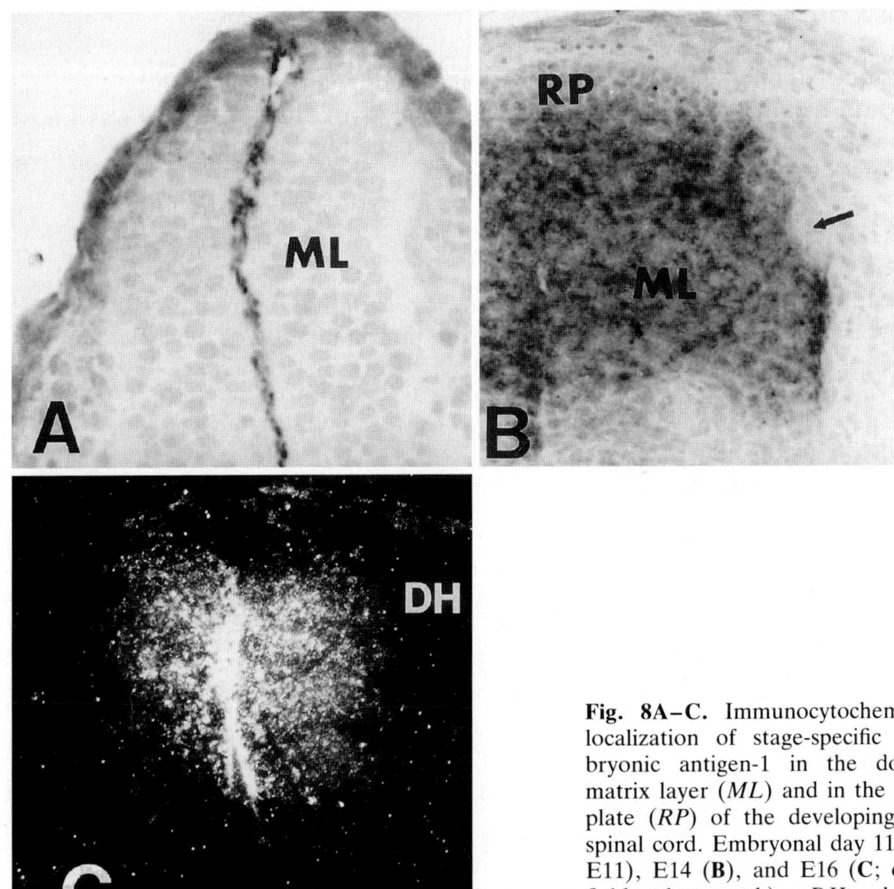

Fig. 8A–C. Immunocytochemical localization of stage-specific embryonic antigen-1 in the dorsal matrix layer (*ML*) and in the roof plate (*RP*) of the developing rat spinal cord. Embryonal day 11 (**A**; E11), E14 (**B**), and E16 (**C**; dark field photograph). *DH*, dorsal horn

vicinity of the spinal cord ventricle (Fig. 8A). At E13, abundant SSEA-1 staining was also detected throughout the alar plate. By this day, the immunoreactivity was found to slightly transgress the level of the sulcus limitans. The antigen, therefore, was clearly present in the intermediate region of the matrix layer. One day later, at E14, SSEA-1 staining was restricted to the alar plate only (Fig. 8B). The staining pattern did not exceed the level of the sulcus limitans. Two days later, intense SSEA-1 staining was still present in the gradually declining alar plate (Fig. 8C). At E17, the antigen appeared to be sparsely located, and at E18 completely abolished from the alar plate.

It should be mentioned that during the described period, SSEA-1 immunoreactivity in the alar plate was mainly located "around" the cells of the alar plate and intermediate region of the matrix layer. The precipitate was often found surrounding the unstained cytoplasma of the cells. During the embryonic and postnatal period of spinal cord development, SSEA-1 immunoreactivity appeared to be absent from the roof plate of the matrix layer (Fig. 8B).

4.2.2.5 Acetylcholinesterase

In the developing and the mature rat spinal cord, AChE could not be detected either in the alar plate or in the roof plate of the matrix layer. This was found to be the case in both the nonfixed cryostat sections and in the fixed gelatin sections.

4.2.2.6 Vimentin

At E12, vimentin was clearly present in the membrana limitans externa (the external border of the matrix layer). In the intermediate area, short, vimentin-

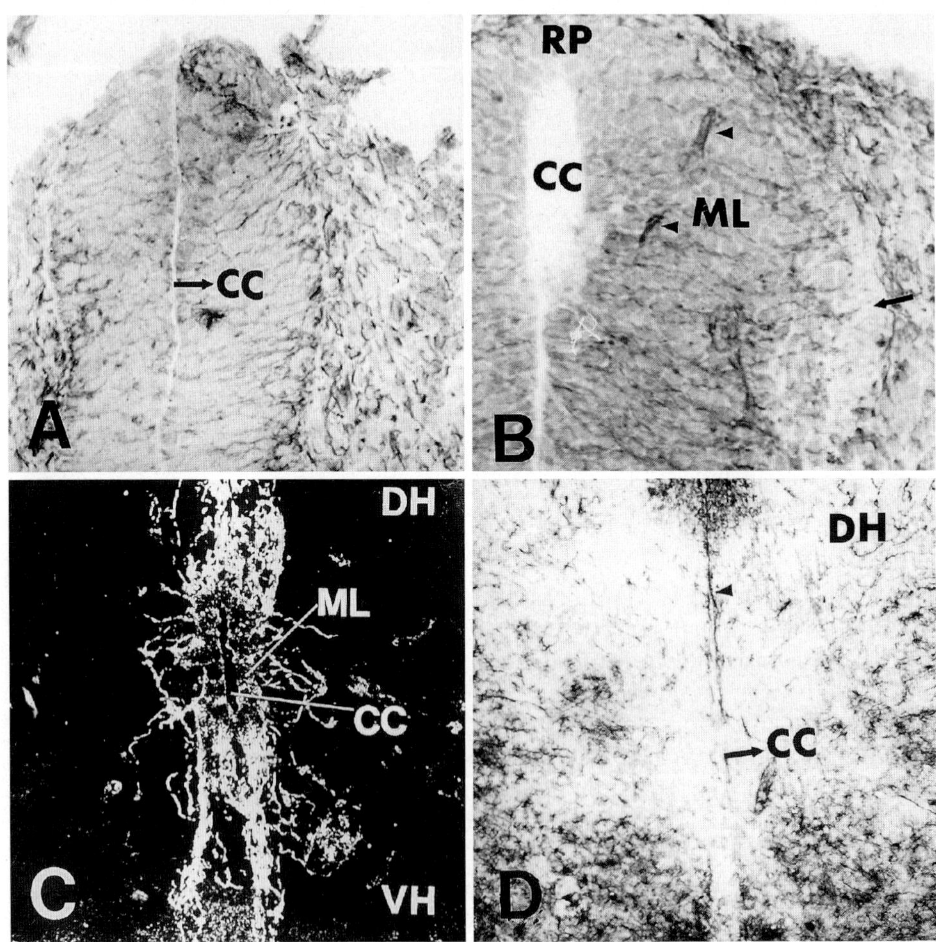

Fig. 9A–D. Vimentin and glial fibrillary acidic protein in the dorsal matrix layer (*ML*) and roof plate (*RP*) of the developing rat spinal cord. Vimentin at embryonal day 12 (**A**; E12), E14 (**B**; *arrowhead* points at vimentin-positive tangle and *large arrow* points at the dorsal root entrance zone), and E20 (**C**) (dark field photograph). **D** Glial fibrillary acidic protein at postnatal day 4 (*arrowhead* points at thin dorsal raphe). *CC*, central canal; *DH*, dorsal horn; *VH*, ventral horn

positive protrusions were found to penetrate from this membrane into the matrix layer. At the same age, a faint scattered pattern of vimentin staining could be detected throughout the alar plate and the intermediate region of the matrix layer. During the following days, this scattered pattern changed into a more regular, radially orientated pattern of intensely stained, vimentin-positive processes (Fig. 9A). The staining intensity was most abundant in the intermediate region of the matrix layer (Fig. 9B). At E13–E14, tangles of short, vimentin-positive fibers were also found to be present in the alar plate and in the intermediate region (Fig. 9B). Most of the short, vimentin-positive fibers emerged from these tangles towards the ventral aspect of the cord. After E17, vimentin-positive fibers appeared to originate from the intermediate region of the matrix layer (Fig. 9C). This staining pattern persisted until the first postnatal week. The vimentin staining intensity, however, markedly decreased during this period towards a faint appearance. At P4, the alar plate as well as the intermediate region appeared to be almost completely devoid of vimentin. Surprisingly, after P10 and especially in the mature rat, short, twisting, vimentin-positive fibers were found to emerge in all directions from the dorsal as well as the intermediate region of the spinal cord matrix layer. These fibers were found to penetrate the mantle layer of the cord.

Between E12–E14, the roof plate of the rat spinal cord demonstrated a few faintly stained, vimentin-positive fibers. At the end of the embryonal period, hardly any vimentin positivity could be detected in the spinal cord roof plate.

4.2.2.7 Glial Fibrillary Acidic Protein

During the embryonal phase of spinal cord development, GFAP immunoreactivity was found to be absent from the alar plate and the intermediate region of the matrix layer, except for a sporadic fiber which traversed the matrix layer. After birth, some GFAP-positive fibers were found to be present in the intermediate regions of the matrix layer in particular. These fibers curved sharply after leaving the matrix layer and formed a thin raphe in the dorsal mantle layer (Fig. 9D; see Sect. 4.3.3.7).

During the embryonal phase as well as the postnatal phase of spinal cord development, GFAP immunoreactivity appeared to be absent from the roof plate of the spinal cord matrix layer.

4.3 Mantle Layer

4.3.1 Ventral Horn/Ventral Commissure/Ventral Raphe

4.3.1.1 Cytoarchitecture

After their last mitotic cycle, the neuroblasts of the ventral matrix layer migrate towards the periphery of the neural tube to form the ventral part of the mantle layer, i.e., the ventral horn. In the ventral horn, the neuroblasts differentiate into the spinal cord motor neurons. This topographical relationship between the matrix and mantle layer was demonstrated by Nornes and Das (1974). The occupation of the ventral horn by the motor neurons starts around E12. The three classes of motor neurons, the larger α-cells and the smaller β- and γ-cells, intermingle and aggregate to form a lateral and medial motor column (note that at the thoracic levels, only one motor cell column can be discerned). The aggregation into the different motor columns takes place between E13 and E17 (Altman and Bayer 1984). The larger α-motor neurons differentiate ahead of the smaller ones (Altman and Bayer 1984). The motor neurons convey the outgoing information by way of their peripheral axons, of which the first can already be seen to leave the spinal cord at E12 (Altman and Bayer 1984; Oudega et al. 1990a). During the embryonal phase of development, the ventral horn appears to be occupied with closely packed, lightly and medium darkly cresyl violet-stained cells. Especially during the early embryonal days, the ventral mantle layer can easily be distinguished from the ventral matrix layer, whereas the latter contains more heavily cresyl violet-stained cells. After birth, the ventral horn cells are found to be more loosely organized (Rexed's Lamina VIII, Rexed 1954), and most of them stain mediumly or lightly for the Nissl substance, with the exception of the motor cells (Rexed's lamina IX; Rexed 1954), which appear as heavily cresyl violet-stained cells.

The ventral commissure is formed by the decussating fibers of the contralaterally projecting relay neurons of the intermediate gray. With cresyl violet staining, only small round cells, most likely glial cells, can be found in this region.

4.3.1.2 Neurofilaments

At E13, the first weak neurofilament immunoreactivity was detected in cells of the incipient ventral horn (Fig. 10A). During the following days of embryonal development, the staining increased in intensity and appeared to be mainly present in the cells of the motor columns (Fig. 10B). Nevertheless, thin immunoreactive fibers were also discerned throughout the developing ventral horn. After P8, neurofilament expression declined towards a faint appearance. At the end of the second postnatal week, neurofilament immunoreactivity was found to be absent from most of the motor neurons (Fig. 10C,D).

At E13, the first faintly stained neurofilament-positive fibers were detected in the ventral commissure at the cervical levels of the cord. At thoracic and lumbar levels, only a single neurofilament-positive fiber was discerned in this region. One day later, the ventral commissure showed a similar, intensely stained fiber pattern at all spinal cord levels. During further development and at maturity, the ventral commissure remained moderately stained for neurofilaments. Throughout the embryonal and postnatal phase of spinal cord development, neurofilament immunoreactivity was found to be absent from the ventral raphe.

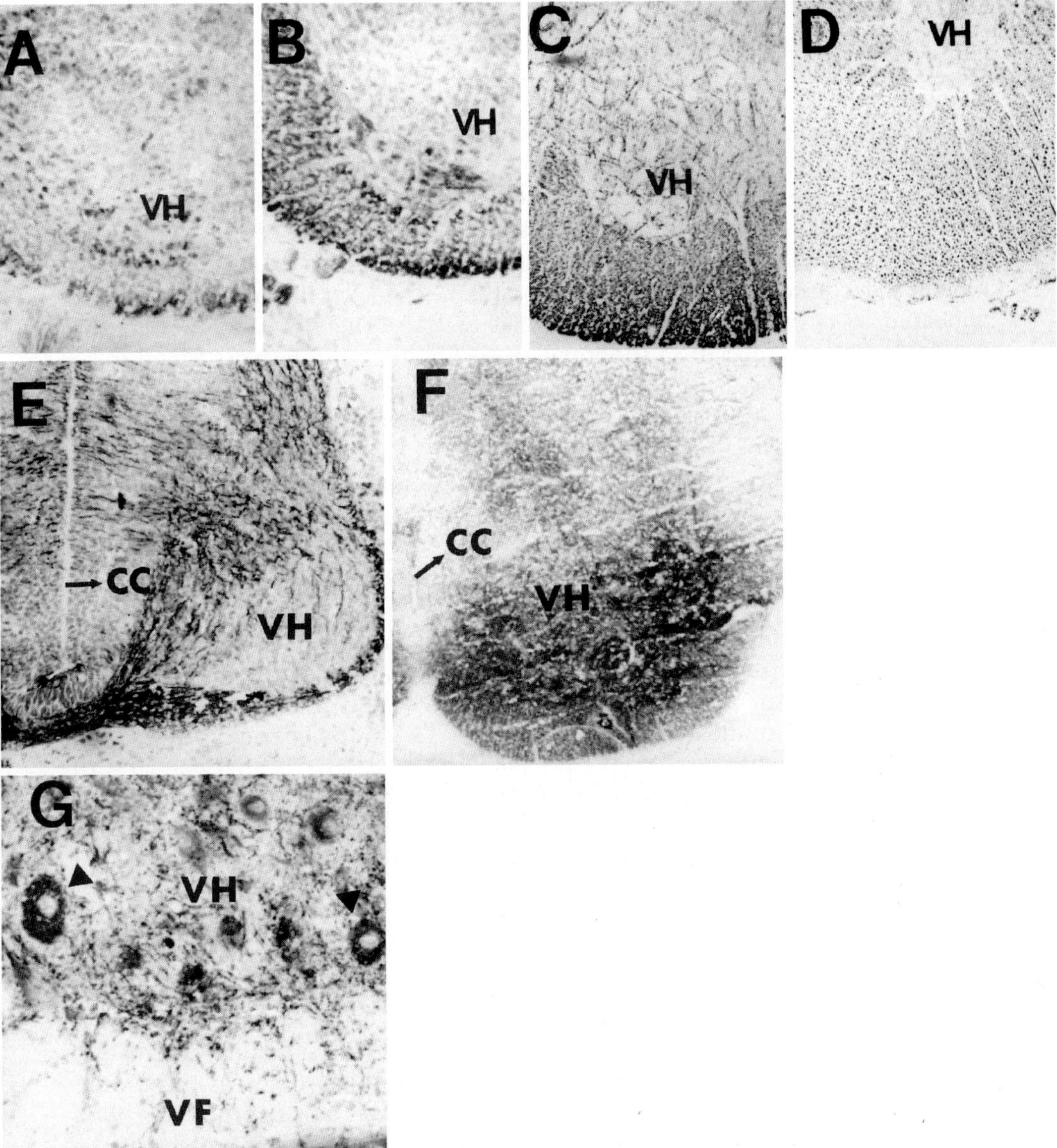

Fig. 10A–G. Immunocytochemical localization of neurofilament and microtubule-associated protein 1A in the developing, ventral rat spinal cord. Neurofilament at embryonal day 13 (**A**; E13), E16 (**B**), postnatal day 8 (**C**), and in adult (**D**). Microtubule-associated protein 1A at E14 (**E**), E18 (**F**), and in adult (**G**); *arrowheads* point to positive motor neurons. *VH*, ventral horn; *CC*, central canal; *VF*, ventral funiculus

4.3.1.3 Microtubule-Associated Proteins

At E12, the borderline between the ventral matrix layer and the incipient ventral horn was intensely stained for MAP 1A. Sagittal sections revealed the presence of the antigen in cells and in their more faintly stained axons. At E13, this staining pattern could still be detected in the border region. By this day, a small population of motor cells and their axons were moderately stained, whereas the neuropil was found to be faintly stained for the MAP 1A antigen. At E14, the staining intensity of the transversely orientated axons in the border region had considerably increased (Fig. 10E). The overall pattern, however, appeared to be essentially similar to that at E13. Thereafter, until P12, the ventral horn and especially a subpopulation of motor neurons were found to be intensely stained for the MAP 1A antigen (Fig. 10F). After P12, the staining intensity gradually decreased in the neuropil, and around P20 the ventral horn demonstrated an overall faint presence of MAP 1A. Surprisingly, from P20 on, some motor neurons in the ventral horn were found to express the MAP 1A antigen (Fig. 10G). As mentioned above, at E13 the axons of the motor neurons appeared to be positive for MAP 1A. These fibers remained positive until maturity. From P12 on, however, their staining intensity gradually decreased towards a faint appearance.

At E14, the first MAP 1A-stained fibers were detected in the ventral commissure of the rat spinal cord (Fig. 10E). Thereafter, this region remained positive for the antigen. After P8, the immunostaining gradually decreased in its intensity, and around P12 hardly any MAP 1A positivity could be detected in the ventral commissure.

During the embryonal phase of development, MAP 1A immunoreactivity appeared to be absent from the ventral raphe of the spinal cord (Fig. 10F). After birth and in the adult rat spinal cord, the ventral raphe remained devoid of MAP 1A positivity.

At E12, the first moderate MAP 2 immunoreactivity was found in the ventral horn. During the following days of spinal cord development, the staining intensity gradually developed into an abundant appearance (Fig. 11A). The antigen was found mainly in the neuropil, but some positive motor neurons could also be discerned (Fig. 11B). After P12, MAP 2 immunostaining decreased towards a moderate intensity and could still be observed as such in the adult ventral horn. From P30 on, a subpopulation of motor neurons remained strongly immunoreactive for the MAP 2 antigen (Fig. 11C). The axons of the ventral motor cells which are gathered in the spinal cord ventral roots appeared to be devoid of MAP 2 immunoreactivity.

Throughout the embryonal (Fig. 11A,B) and postnatal development (Fig. 11C) and in the adult rat spinal cord, MAP 2 immunoreactivity appeared to be absent from the ventral commissure and the ventral raphe.

Fig. 11A–F. Microtubule-associated proteins 2 and 5 in the developing ventral rat spinal cord. Microtubule-associated protein 2 at embryonal day 13 (**A**; E13), E18 (**B**), and postnatal day 20 (**C**). Microtubule-associated protein 5 at E13 (**D**), E18 (**E**), and in adult (**F**). *ML*, matrix layer; *VH*, ventral horn; *IG*, intermediate gray; *VF*, ventral funiculus

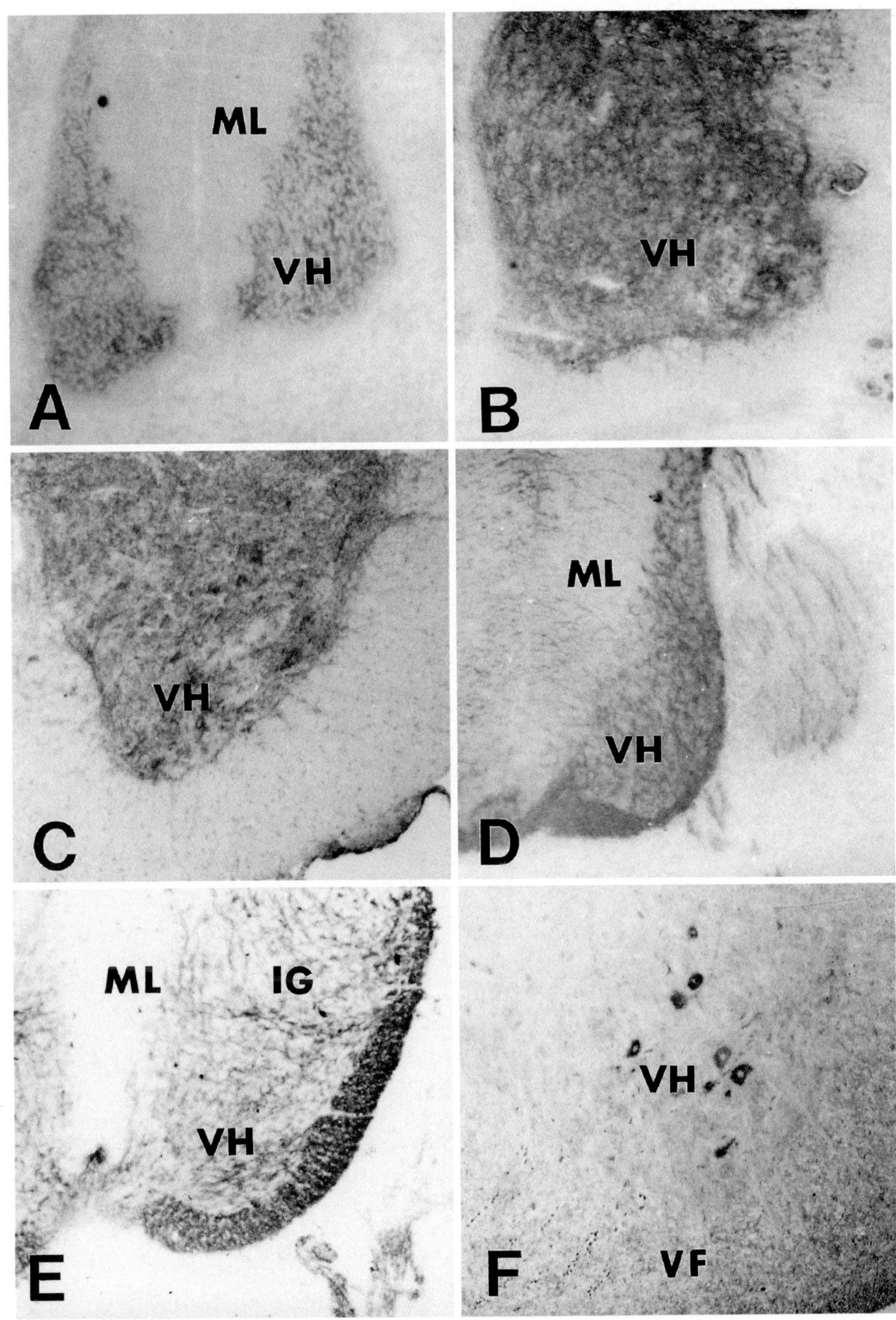

ML
VH
A
VH
B
VH
C
ML
VH
D
ML
IG
VH
E
VH
VF
F

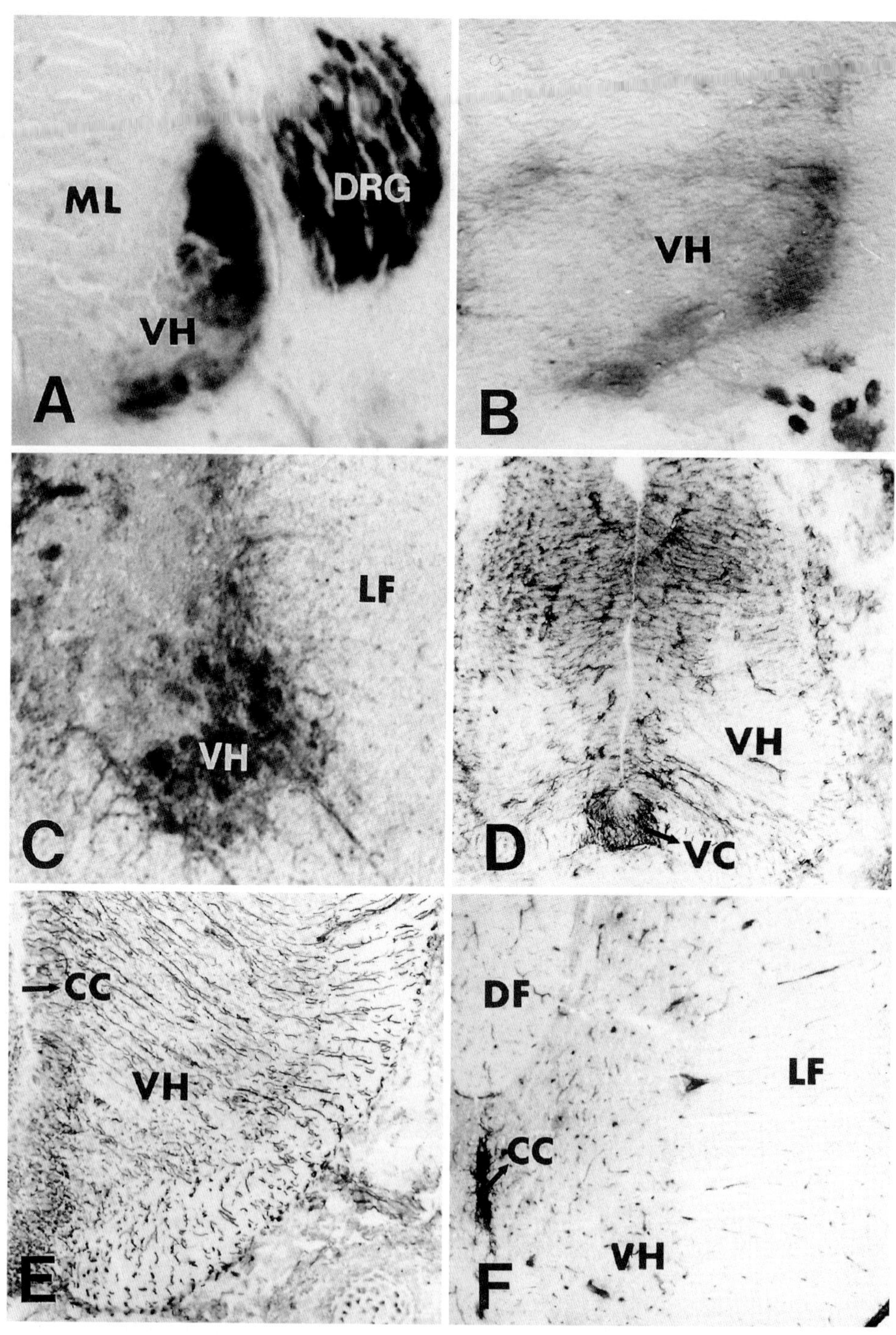
ML
DRG
VH
A
VH
B
LF
VH
C
VH
VC
D
CC
VH
E
DF
LF
CC
VH
F

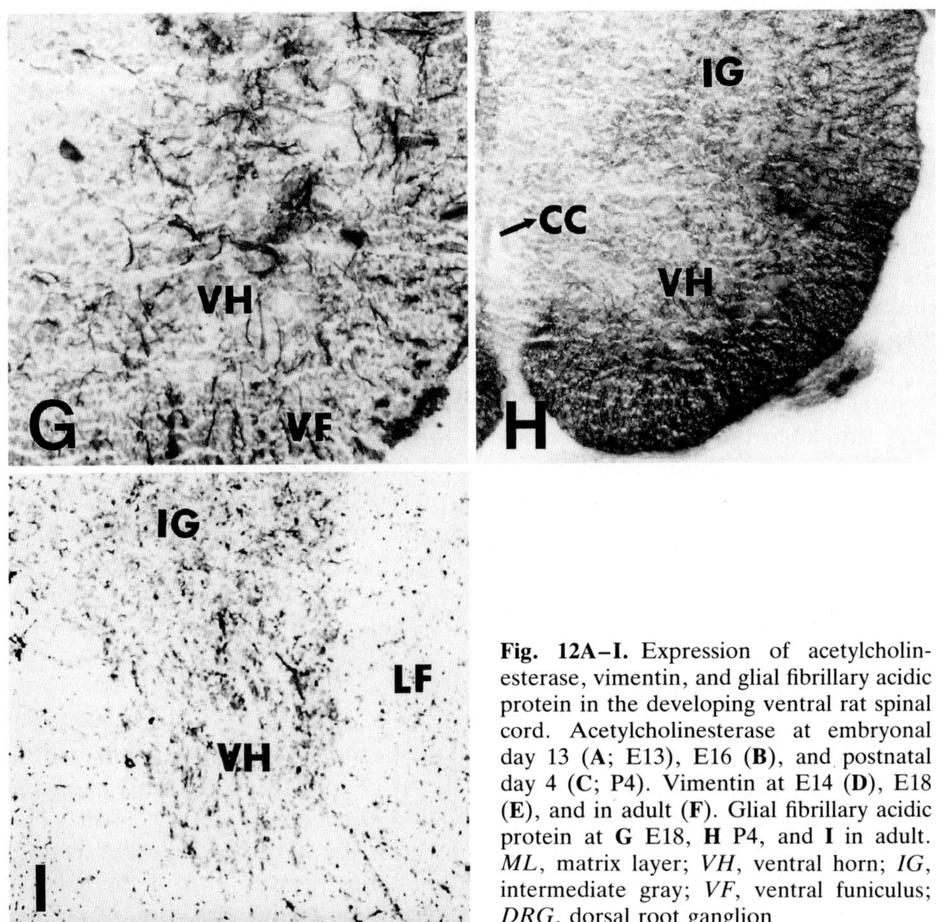

Fig. 12A–I. Expression of acetylcholinesterase, vimentin, and glial fibrillary acidic protein in the developing ventral rat spinal cord. Acetylcholinesterase at embryonal day 13 (**A**; E13), E16 (**B**), and postnatal day 4 (**C**; P4). Vimentin at E14 (**D**), E18 (**E**), and in adult (**F**). Glial fibrillary acidic protein at **G** E18, **H** P4, and **I** in adult. *ML*, matrix layer; *VH*, ventral horn; *IG*, intermediate gray; *VF*, ventral funiculus; *DRG*, dorsal root ganglion

At E13, the developing ventral horn abundantly expressed the MAP 5 antigen (Fig. 11D). Immunoreactivity was mainly detected in the neuropil, but some positive motor neurons were also present. Additionally, MAP 5-positive fibers were also detected in the ventral roots (Fig. 11D). During the following 4 days, the staining pattern remained essentially similar, but its intensity substantially increased towards an abundant appearance. Thereafter, from E18 until P10, the MAP 5 staining pattern and intensity remained essentially the same (Fig. 11E). From P10 on, the reactivity gradually decreased towards a faintly stained ventral horn at maturity. The mature staining pattern appeared to be present around P20. Surprisingly, at this age, a subpopulation of motor neurons remained positive for the MAP 5 antigen (Fig. 11F). The axons of the adult spinal cord motor neurons were found to be faintly stained for the MAP 5 antigen.

At E13, the ventral commissure demonstrated the first MAP 5-positive fibers (Fig. 11D). From this age on until P10, the ventral commissure remained positive, although its staining never appeared to be intense (Fig. 11E). After

P10, MAP 5 immunoreactivity gradually decreased towards a faint staining at maturity. Throughout the development of the spinal cord as well as in the adult cord, the ventral raphe was found to be devoided of MAP 5 immunoreactivity.

4.3.1.4 Stage-Specific Embryonic Antigen-1

During the embryonal phase of spinal cord development, SSEA-1 appeared to be absent from the ventral horn. After birth, however, faint, coarse SSEA-1 staining was discerned. This type of SSEA-1 immunoreactivity could still be detected in the ventral horn of the adult spinal cord.

During the embryonal period, SSEA-1 was found to be absent from the ventral commissure. After birth, this region developed a coarse staining pattern similar to the one described above for the ventral horn. At maturity, the commissure still demonstrated a coarse SSEA-1 staining pattern.

From E16 on until maturity, the ventral raphe stained positively for SSEA-1. The intensity of the staining was greatest during the embryonal period. During the postnatal period of development, the ventral raphe was found to be moderately stained for SSEA-1.

4.3.1.5 Acetylcholinesterase

At E12, a patch-like AChE staining could be found in the incipient ventral horn. This pattern seemed to be caused by AChE-positive cells, most likely the first spinal cord motor neurons to settle in the ventral mantle layer. This distribution pattern was demonstrated in both the nonfixed cryostat sections and the fixed gelatin sections. One day later, the ventral horn of the cervical and lumbar levels demonstrated a heterogeneous AChE distribution pattern (Fig. 12A). Two cell concentrations, most likely representing the future ventral and medial motor cell columns, appeared to be abundantly stained for the enzyme. The neuropil at this time demonstrated faint AChE staining. The two cell clusters seemed not to be completely separated from each other. Usually, at the thoracic levels only one such AChE-positive cell cluster is detected. At E13, the ventral motor roots were also found to express the enzyme (Fig. 12A). Three days later, at E16, high AChE expression could be discerned in the cells of the motor columns and in their emerging axons (Fig. 12B). By this day, at the cervical and lumbar levels the two cell groups were found separately from each other. At E18, the ventral motor columns demonstrated abundant AChE staining, whereas the neuropil appeared to be moderately stained. During the following days of embryonal development and in the first few postnatal days, the distribution pattern remained essentially similar. Thereafter, the ventral horn gradually developed a moderate, homogeneously distributed AChE staining pattern. The activity was mainly found in the ventral horn neuropil, although clearly stained motor neurons could still be detected (Fig. 12C).

During postnatal development, the location and size of the AChE-positive motor cell columns appeared to depend on the spinal cord level examined. At the cervical and lumbar levels, two distinct groups could be discerned, medially and laterally in the ventral horn. At the thoracic level, usually one cell group was discerned, although in some cases two smaller, partially overlapping cell groups were also found. After P4, the cells in the motor columns seemed to enlarge and were found to demonstrate a clear AChE activity in their cytoplasm (Fig. 12C). After P12, the ventral horn also contained small, isolated, AChE-positive cells, whereas faint staining was found to be present throughout the horn. From P20 on, the adult distribution pattern of AChE was detected in the ventral horn of the spinal cord. At P20, isolated AChE-positive cells were scattered throughout the ventral horn. The motor cell column clearly stained positively for AChE. The ventral roots, however, lacked the enzyme. AChE-positive dendrites from ventral horn cells were found to form an extensive "network" in the funiculi. During the embryonal and postnatal development of the rat spinal cord, the ventral commissure and the ventral raphe appeared to be devoid of AChE.

4.3.1.6 Vimentin

At E13, the first vimentin immunoreactivity appeared to be present in the ventral horn (Fig. 12D). The vimentin positivity was demonstrated as an intense radial staining pattern. The fibers seemed to originate from the matrix layer (see Sect. 4.2.1) and following an oblique course towards the ventro-lateral funiculus. During the following 3 days, the ventral horn developed an abundant radial pattern of vimentin-positive fibers. Between E16 and E22, the staining pattern and intensity remained essentially similar (Fig. 12E). After birth, the staining intensity gradually decreased towards a faint appearance of vimentin in the ventral horn. After P10, vimentin immunoreactivity could only be found scattered throughout the ventral horn, mostly in cells, although single, short, positive protrusions were also present. Finally, at P20, the adult vimentin distribution pattern could be discerned, demonstrating only a single vimentin-immunoreactive, short fiber in the ventral horn. In the vicinity of the ventral aspect of the central canal, however, several twisting, thin, vimentin-positive fibers were found to penetrate from the matrix layer into the ventral part of the mantle layer (Fig. 12F).

At E16, the ventral commissure contained a single fiber-like, vimentin-immunoreactive structure. After E18, however, this region appeared to be invaded by vimentin-positive fibers.

At E16, the ventral raphe was found to be intensely stained for vimentin. The immunoreactive fibers appeared to originate in the ventral part of the matrix layer and, after a sharp curve, to course in a vertical direction to establish a ventral raphe (see Sect. 4.2.1.6). From E18 on, the vimentin staining remained intense, but after P10 it decreased towards a faint appearance. From P20 on, the ventral raphe was found to contain only a few vimentin-positive fibers.

4.3.1.7 Glial Fibrillary Acidic Protein

At E16, some faintly stained GFAP-immunoreactive protrusions were present scattered throughout the ventral horn. Two days later, at E18, the GFAP antigen was found concentrated in the ventrolateral region of the horn (Fig. 12G). These fibers were thin and diffusely scattered over this region. A similar pattern was also detected between E19 and E22, during which period the staining intensity considerably increased. At P2, the ventral horn appeared to demonstrate radially orientated GFAP-positive fibers. The fibers emerged from the matrix layer and diverged towards the peripheral white matter. Between P4 and P10, the GFAP immunostaining increased dramatically (Fig. 12H). Throughout the ventral horn, short, positive protrusions could also be seen. At P20, the adult staining pattern was found in the ventral horn. GFAP-positive cells, with several short protrusions, were located throughout the ventral mantle layer (Fig. 12I). Most likely, this particular GFAP staining pattern reflected the presence of the stellate astrocytes in the adult spinal cord.

After birth, the ventral commissure gradually developed GFAP positivity. Most of the GFAP-immunoreactive fibers were radially arranged in the commissure. In the adult ventral commissure, only a few GFAP-positive fibers could still be detected.

During the first few postnatal weeks, the ventral raphe gradually developed intense GFAP immunoreactivity. The GFAP-positive fibers of the raphe originated from the intermediate region of the matrix layer (see Sect. 4.2.2.7). At P20 and in the adult rat spinal cord, the ventral raphe appeared to contain thin, intensely stained, GFAP-positive fibers.

4.3.2 Intermediate Gray

4.3.2.1 Cytoarchitecture

After their last mitotic cycle, the neuroblasts of the intermediate matrix layer migrate towards the periphery of the neural tube to form the intermediate region of the mantle layer, i.e., the intermediate gray. A topographical relationship between the intermediate regions of the spinal cord matrix and mantle layer has been suggested by Altman and Bayer (1984). Around E13, the relay cells start to occupy the intermediate gray. The neurons arrange themselves in Rexed's laminae IV–VIII, and a small portion appears to be located in Rexed's lamina I (Rexed 1954). The size and form of the cells as well as their density was found to be largely variable in the different laminae of the intermediate gray (Rexed 1954; Altman and Bayer 1984).

During the embryonal period, the intermediate gray was found to be occupied by a heterogeneous group of cresyl violet-stained cells. Darkly, medium darkly, and lightly stained cells can be recognized in the different laminae. Especially during the early embryonal days, the incipient intermediate gray can easily be distinguished from the intermediate area of the matrix layer, since the former is predominantly occupied by more lightly cresyl violet-stained cells. In the adult rat spinal cord, Rexed's lamina V contains Clarke's nucleus (Waibl 1973). At the thoracic levels, lamina VII encloses the intermediolateral cell column in its most lateral aspect. The intermediomedial cell column was found to be present in the medial aspect of lamina VII, just dorsal to the central canal (Waibl 1973; Molander et al. 1984).

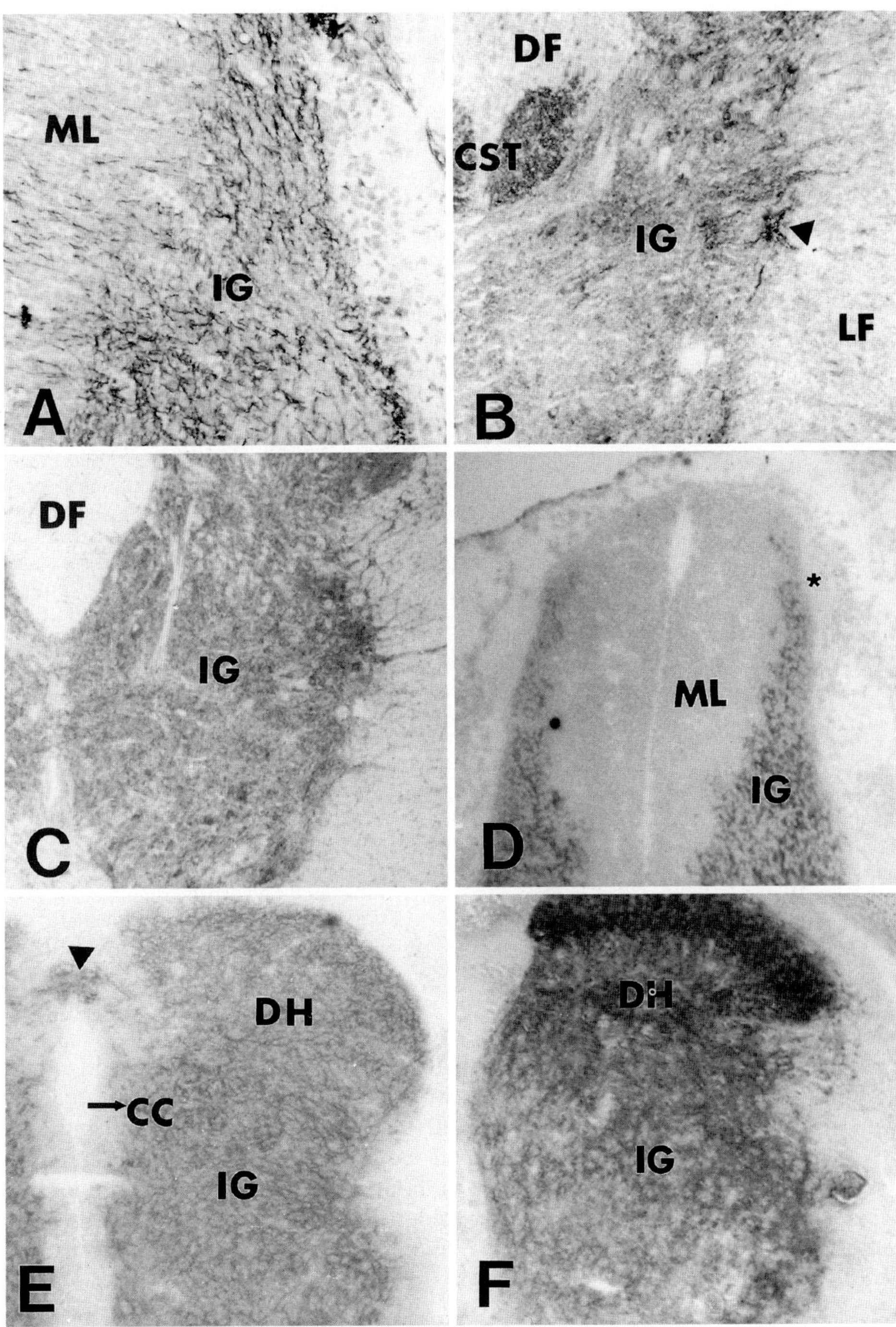

ML
IG
A
DF
CST
IG
LF
B
DF
IG
C
ML
*
IG
D
DH
CC
IG
E
DH
IG
F

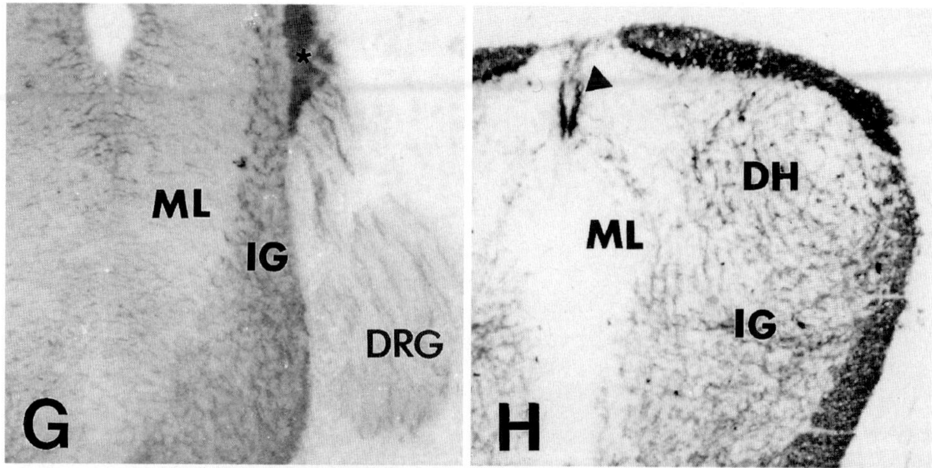

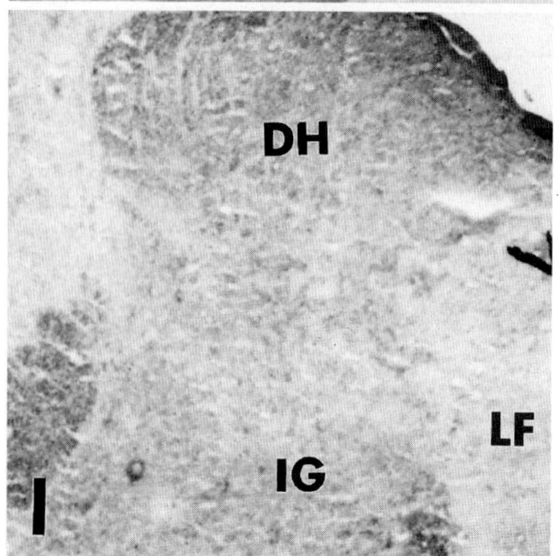

Fig. 13A–I. Immunocytochemical localization of microtubule-associated proteins 1A, 2, and 5 in the developing intermediate gray (*IG*) of the rat spinal cord. Microtubule-associated protein 1A at embryonal day 14 (**A**) and postnatal day 30 (**B**); *arrowhead* points to intermediolateral cell column. Microtubule-associated protein 2 in adult (**C**) and at E13 (**D**; *star* placed in dorsal root entrance zone), E16 (**E**; *arrowhead* points to roof plate), and E18 (**F**). Microtubule-associated protein 5 at E13 (**G**), E16 (**H**; *arrowhead* points to roof plate and dorsal raphe), and in adult (**I**). *ML*, matrix layer; *DF*, dorsal funiculus; *CST*, corticospinal tract; *DH*, dorsal horn; *CC*, central canal; *LF*, lateral funiculus; *DRG*, dorsal root ganglion

4.3.2.2 Neurofilaments

From E16 on, faintly stained neurofilament-positive cells were found in the intermediate gray of the rat spinal cord. In addition, thin immunoreactive fibers could be discerned in this region. Until P10, the pattern and intensity of the neurofilament staining remained essentially similar, but thereafter its intensity gradually declined towards a faint appearance. After P20, only a sparse neurofilament-positive fiber could still be demonstrated, whereas the intermediate mantle layer appeared to be devoid of neurofilament-positive cells.

4.3.2.3 Microtubule-Associated Proteins

At E13, abundantly stained MAP 1A-positive cells were found along the borderline of the matrix layer and the developing intermediate mantle layer.

Sagittal sections also revealed the presence of faintly stained axons in this border region. One day later, additional faint MAP 1A immunoreactivity was detected in the neuropil of the intermediate mantle layer (Fig. 13A). Some scattered, weakly stained cells could also be discerned. During further development, the staining intensity gradually decreased towards a faint appearance and after P10, MAP 1A-positive cells could hardly be detected in the intermediate gray. MAP 1A-positive fibers, however, were still present in the intermediate part of the spinal cord mantle layer (Fig. 13B).

At E13, a small rim of MAP 2 immunoreactivity was discerned between the intermediate matrix layer and the lateral funiculus (Fig. 13D). This region most likely represented the incipient intermediate mantle layer. During further development, MAP 2 was abundantly expressed throughout the intermediate part of the spinal cord mantle layer (Fig. 13E). Single MAP 2-positive cells could also be discerned. After an initial increase (Fig. 13F), the MAP 2 staining intensity gradually decreased after P10. Around P20, the adult overall moderate MAP 2 staining appeared to be present in the intermediate gray (Fig. 13C).

At E13, the developing intermediate mantle layer was found to moderately express the MAP 5 antigen (Fig. 13G). During the next 2 days, the staining pattern remained essentially similar, but its intensity substantially increased. Around E16, MAP 5-positive fibers traversing the intermediate mantle layer were detected (Fig. 13H). After E18, the staining intensity decreased towards a faint appearance around P10 and at maturity, only faint MAP 5 staining could still be discerned (Fig. 13I). Additionally, at the thoracic spinal cord levels the lateral horn, which is known to enclose the intermediolateral cell column (Molander et al. 1984), was found to contain MAP 5-positive cells (Fig. 13I).

4.3.2.4 Stage-Specific Embryonic Antigen-1

During the embryonal phase of spinal cord development, the intermediate mantle layer appeared to be devoid of SSEA-1 immunoreactivity. After birth, this region gradually developed faint, coarse SSEA-1 staining. The activity was found to be distributed throughout the intermediate gray and could still be detected in the adult rat spinal cord.

4.3.2.5 Acetylcholinesterase

At E14, faint AChE expression was detected throughout the intermediate region of the mantle layer. One day later, the enzyme was found in cells of the lateral region of the intermediate mantle layer (Fig. 14A). Most likely, these cells represented the developing intermediolateral cell column. Medial to this cell column and close to the sulcus limitans, another AChE-positive region could be discerned. This cell group probably corresponded to the developing intermediomedial cell column. The development of these autonomic cell groups will be dealt with separately (see Sect. 4.3.2.6). Between E15 and P12, the developing intermediate part of the spinal cord mantle layer was found to

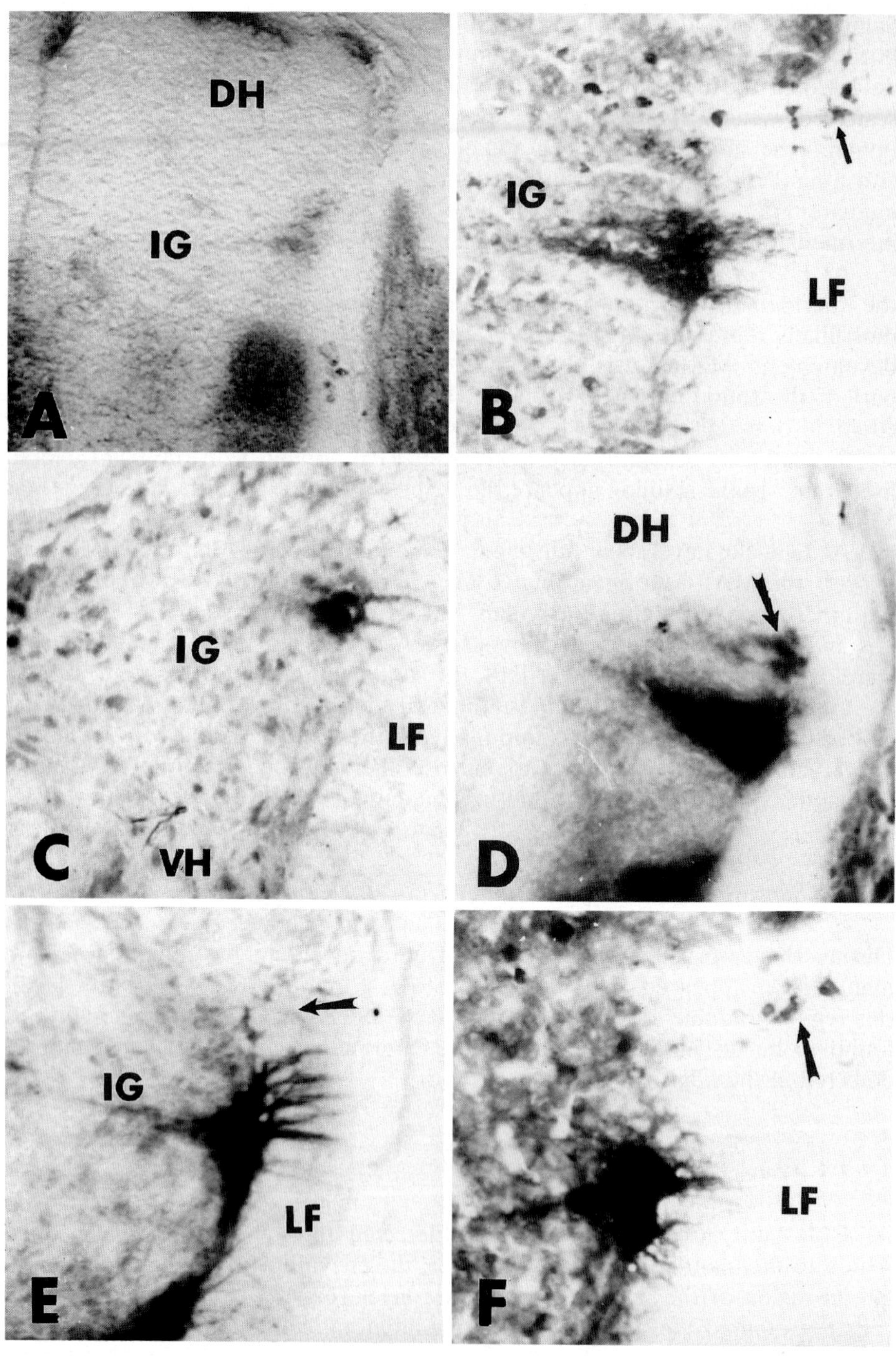
DH
IG
A
IG
LF
B
IG
LF
C
VH
DH
D
IG
LF
E
LF
F

persist in its faint AChE expression (Fig. 14B). After P12, isolated AChE-positive cells could be discerned in the intermediate gray. After E20 and in the adult cord, the distribution pattern of AChE appeared to be essentially similar, with numerous AChE-positive cells embedded in the faintly stained neuropil of the intermediate gray (Fig. 14C).

4.3.2.6 The Autonomic System

At E15, AChE was observed in cells of the lateral region of the intermediate part of the mantle layer (Fig. 14A). Most likely, these cells corresponded to the developing intermediolateral cell column, which is situated in the lateral horn of the adult rat (see Molander et al. 1984). The positive cells were clearly visible at the thoracic levels, but absent at the cervical levels. In the lumbar spinal cord, however, some AChE-positive cells could also be detected in this particular region. One day later, at E16, the enzyme appeared to be more intensely expressed in the lateral horn cells (Fig. 14D). In some cases, more medially positioned cells close to the sulcus limitans in the thoracic cord also stained positively for the enzyme. Most likely, this cell group represented the developing intermediomedial cell column, which is found dorsolateral to the central canal in the adult rat spinal cord (Molander et al. 1984). At E18, after the shrinkage of the central canal has started, the medial cell group indeed appeared to be located dorsolaterally to the spinal cord ventricle. Between the two cell groups, an AChE-positive area was also demonstrated. At the light microscopical level, however, it could not be determined whether in this region the enzyme was located in cells or in fibers. In addition on these three positive regions, AChE was found to be present in cells in the dorsal part of the lateral funiculus, just beneath the dorsal horn. Between E18 and E20, the staining intensity in the lateral and medial cell group was found to increase (Fig. 14E). In addition, AChE-positive fibers were found to emerge into the lateral funiculus originating from the cells of the intermediolateral column (Fig. 14E). These fibers were found to traverse the lateral funiculus. During the first few postnatal days, the laterally and medially located AChE-positive cell group could be studied more closely in the cresyl violet-stained sections and could be definitely identified as the intermediolateral and intermediomedial cell column, respectively (Fig. 14F). From P12 on, AChE diminished in the area between the two cell groups as well as in the cells of the intermediomedial column. After P16, the enzyme also disappeared from the fibers which emerged from the cells of the intermediolateral column. Throughout postnatal development, AChE-positive cells were found to be present in the dorsal aspect of the lateral funiculus. The mature staining pattern was found to be present around P20,

---

Fig. 14A–F. Acetylcholinesterase in the developing intermediate gray (*IG*) of the rat spinal cord. Embryonal day 15 (**A**; E15), postnatal day 12 (**B**; P12; *arrow* points to positive cells in dorsal part of the lateral funiculus, *LF*), adult (**C**), E16 (**D**; *arrow* points to intermediolateral cell column), E18 (**E**; *arrow* points at positive cells in dorsal LF), and P10 (**F**; *arrow* points at positive cells in dorsal part of the LF). *DII*, dorsal horn; *VH*, ventral horn

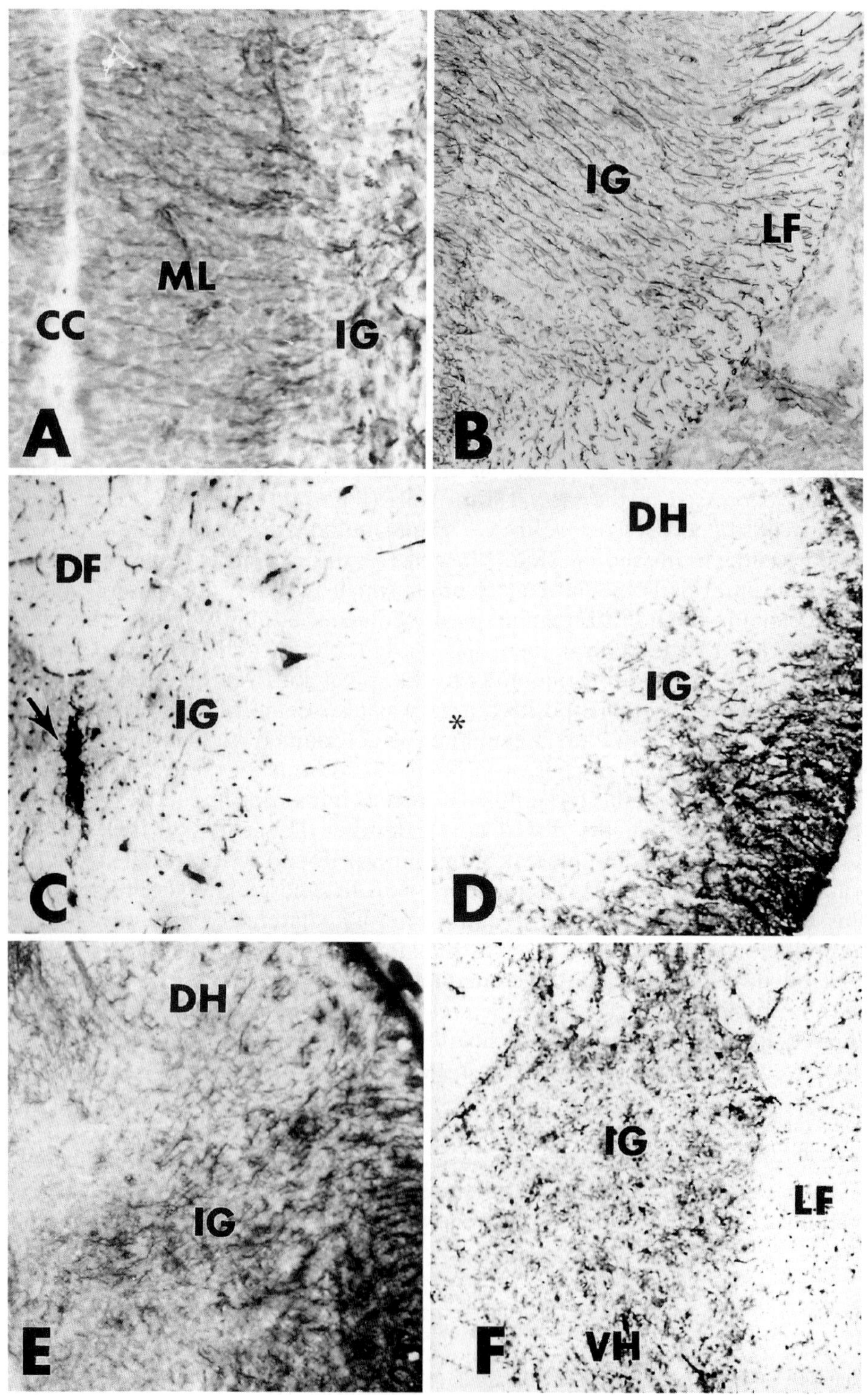

ML
CC
IG
A
IG
LF
B
DF
IG
C
DH
IG
*
D
DH
IG
E
DH
IG
LF
VH
F

with AChE-positive cells in the intermediolateral column as well as in the dorsal part of the lateral funiculus.

4.3.2.7 Vimentin

At E13, the first vimentin immunoreactivity was found in the incipient intermediate mantle layer. The intense staining appeared to be distributed in a radial fiber pattern (Fig. 15A). The fibers seemed to originate from the matrix layer (see Sect. 4.2.2.6). During the following days of embryonal development, the intermediate region of the mantle layer continued to demonstrate such an abundant vimentin-positive radial fiber pattern (Fig. 15B). After birth, the overall staining intensity gradually decreased, and from P10 on only faint vimentin staining was found throughout the intermediate mantle layer. By this time, most of the vimentin staining was found to be present in cells, although single short immunoreactive fibers were also detected. Around P20, the adult distribution pattern was found to be present, with only a single vimentin-immunoreactive fiber in the intermediate part of the spinal cord mantle layer. In the vicinity of the central canal, however, several twisting, thin, vimentin-positive fibers were found to penetrate from the intermediate matrix layer into the intermediate gray (Fig. 15C).

4.3.2.8 Glial Fibrillary Acidic Protein

At E16, weakly stained GFAP-immunoreactive fibers were found to be scattered throughout the intermediate gray. Until birth, a similar staining pattern remained present in the intermediate mantle layer. The general staining intensity substantially increased during the last week of embryonal development. Around E20, additional radially orientated, GFAP-positive fibers were detected in the intermediate mantle layer (Fig. 15D). After birth, around P2, the number of such radially orientated, GFAP-positive fibers had considerably increased. They were found to emerge from the matrix layer (see Sect. 4.2.2.7) and diverged towards the peripheral marginal layer. Subsequently, between P4 and P10, the intensity of the GFAP immunostaining increased dramatically (Fig. 15E). Besides the described radial pattern, short GFAP-positive protrusions were also discerned scattered throughout the intermediate gray. Around P20, the adult staining pattern was found to be present, with many GFAP-positive cells with several short, positive fibers scattered throughout the intermediate gray (Fig. 15F). Most likely, this distinct staining pattern reflected the presence of the stellate astrocytes of the adult spinal cord intermediate mantle layer.

◄───

Fig. 15A–F. Immunocytochemical localization of vimentin (**A–C**) and glial fibrillary acidic protein (**D–F**) in the intermediate gray (*IG*) of the developing rat spinal cord. Embryonal day 13 (**A**; E13), E18 (**B**), adult (**C**), E20 (**D**; *star* is placed in central canal, *CC*), postnatal day 12 (**E** P12), and P20 (**F**). *ML*, matrix layer; *LF*, lateral funiculus; *DF*, dorsal funiculus; *DII*, dorsal horn; *VH*, ventral horn

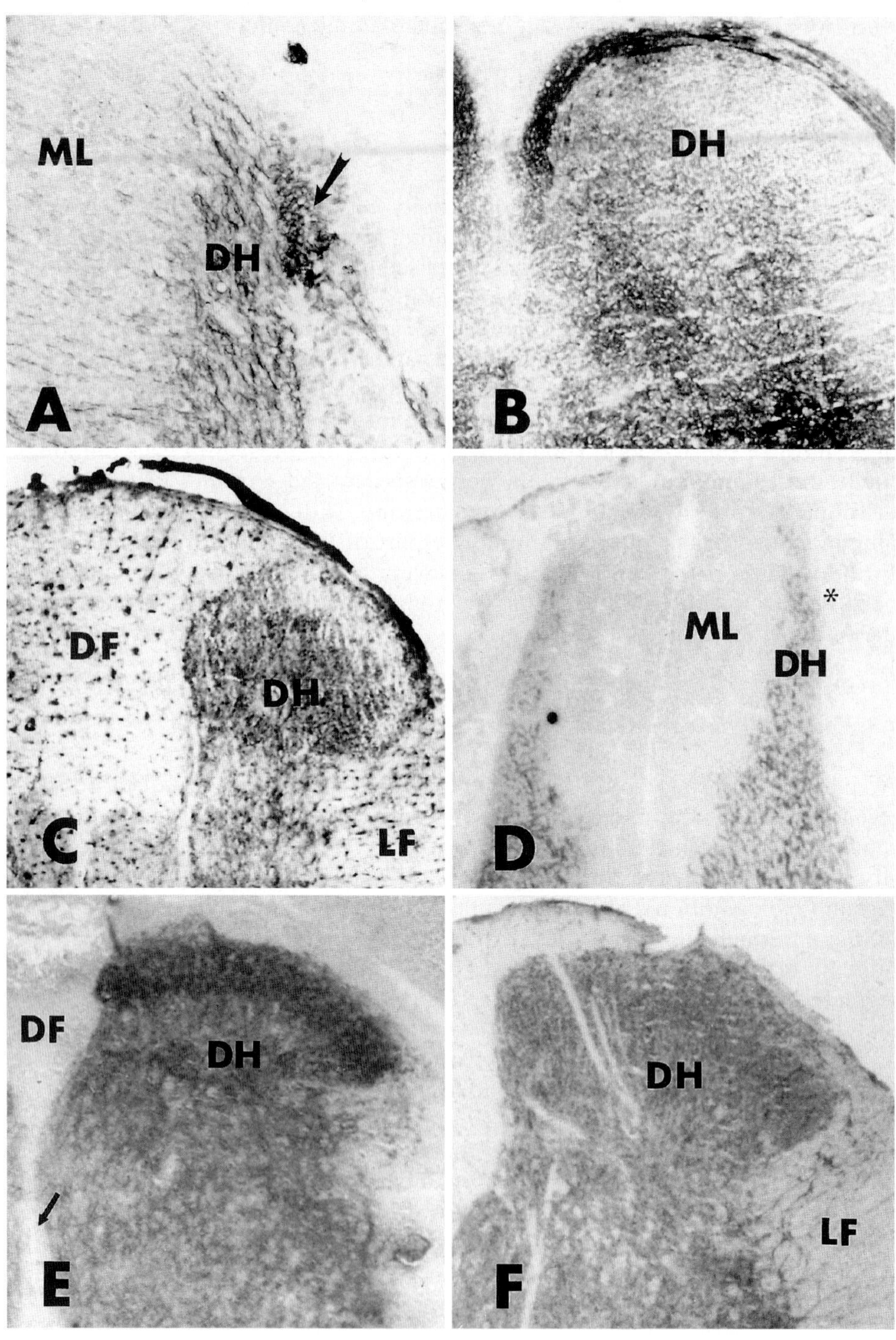
ML
DH
A
DH
B
DF
DH
C
LF
ML
DH
*
D
DF
DH
E
DH
F
LF

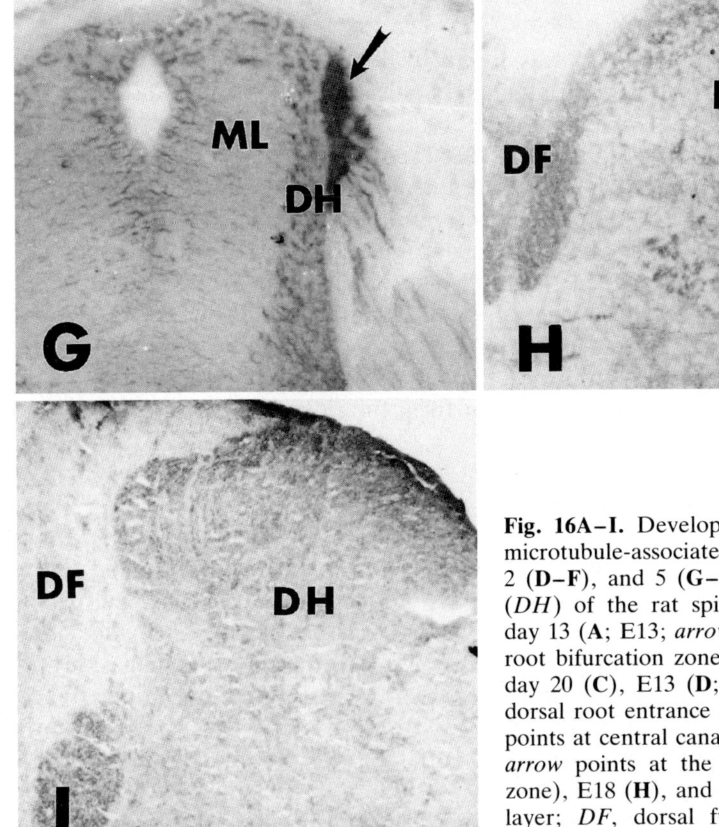

Fig. 16A–I. Developmental expression of microtubule-associated proteins 1A (**A–C**), 2 (**D–F**), and 5 (**G–I**) in the dorsal horn (*DH*) of the rat spinal cord. Embryonal day 13 (**A**; E13; *arrow* points at the dorsal root bifurcation zone), E18 (**B**), postnatal day 20 (**C**), E13 (**D**; *star* is placed in the dorsal root entrance zone), E18 (**E**; *arrow* points at central canal), adult (**F**), E13 (**G**; *arrow* points at the dorsal root entrance zone), E18 (**H**), and adult (**I**). *ML*, matrix layer; *DF*, dorsal funiculus; *LF*, lateral funiculus

4.3.3 Dorsal Horn/Dorsal Raphe

4.3.3.1 Cytoarchitecture

After their last mitotic cycle, the neuroblasts of the dorsal matrix layer (alar plate) migrate towards the periphery of the neural tube to form the dorsal mantle layer, i.e., the dorsal horn. A topographical relationship between the spinal cord dorsal matrix and dorsal mantle layer has been demonstrated by Nornes and Das (1974). The occupation of the dorsal horn by the interneurons starts around E14. The cells arrange themselves in Rexed's laminae II and III (Rexed 1954).

During the early embryonal days, the incipient dorsal horn can easily be distinguished from the alar plate, since the former is occupied by more lightly cresyl violet-stained cells. In the adult rat spinal cord, laminae II and III are occupied by a large variety of cells (see Brown 1981). Lamina II predominantly contains small, darkly cresyl violet-stained cells. Lamina III was found to enclose a heterogeneous group of Nissl-stained cells which appeared to be less closely packed than those in lamina II.

4.3.3.2 Neurofilaments

From E16 on, the dorsal horn was found to demonstrate faint neurofilament immunoreactivity. Mostly, neurofilaments could be detected in thin fibers, but some faintly positive cells were also discerned. Most striking, however, were the intensely stained fibers that entered at the dorsomedial aspect and coursed towards the so-called neck region of the dorsal mantle layer. Most likely, these fibers represent the incoming primary afferents. After birth, the neurofilament immunoreactivity gradually declined and around P12 hardly any positive cells or fibers could be discerned in the dorsal horn. However, in the adult spinal cord, the fibers that course between the dorsal funiculus and the neck region of the dorsal horn remained detectable with neurofilament antibodies. During the embryonal and the postnatal period of spinal cord development, neurofilament immunoreactivity appeared to be absent from the dorsal raphe.

4.3.3.3 Microtubule-Associated Proteins

At E13, the border area between the matrix layer and the developing dorsal horn was found to exhibit the MAP 1A antigen (Fig. 16A). Most of the immunoreactivity was located in the dorsal horn neuropil, although single MAP 1A-positive cells could also be discerned. During the following 3 days of development, the staining intensity increased towards a moderate appearance. Thereafter, starting around E17, the MAP 1A intensity gradually declined towards faint staining at maturity (Fig. 16B). Within the faintly stained dorsal horn, the superficial layers, which could be identified in adjacent cresyl violet-stained sections as Rexed's laminae I and II, lacked the MAP 1A antigen (Fig. 16C). During embryonal and postnatal development, the dorsal raphe of the spinal cord appeared to be devoid of MAP 1A immunoreactivity.

At E13, thin fibers in the dorsal horn were found to express the MAP 2 antigen (Fig. 16D). During further development, the staining intensity gradually increased towards an intense appearance. MAP 2 was found to be mainly present in the dorsal horn neuropil. The superficial region of the dorsal horn in particular, which could be identified as Rexed's laminae I and II in adjacent cresyl violet-stained sections, demonstrated intense MAP 2 staining (Fig. 16E). After P12, the staining intensity became less abundant throughout the dorsal horn. Nevertheless, the adult rat spinal cord still exhibited moderate MAP 2 staining within the dorsal mantle layer (Fig. 16F). Throughout the embryonal and postnatal period of development, MAP 2 staining was found to be absent from the dorsal raphe of the spinal cord.

At E13, a small rim between the dorsal matrix layer and the lateral funiculus was found to express the MAP 5 antigen (Fig. 16G). This region represented the incipient dorsal horn. During the next 2 days, the staining pattern remained essentially similar, but its intensity substantially increased during this period. At E16, MAP 5-positive fibers were found to be present just dorsal to the central canal, decussating to the contralateral dorsal horn. Between E18 and P10, the MAP 5 staining pattern remained similar in the dorsal horn (Fig. 16H). After P10, however, the intensity gradually decreased towards a faint appearance of the antigen in the mature dorsal horn (Fig. 16I). During the embryonal and postnatal period of development, the dorsal raphe was found to be devoid of MAP 5 immunoreactivity.

4.3.3.4 Stage-Specific Embryonic Antigen-1

At E18, SSEA-1 was found to be moderately expressed in the dorsal horn of the rat spinal cord (Fig. 17A). The antigen appeared to be present in the extracellular space. By this time, SSEA-1 staining could be detected in three different regions within the dorsal horn. These areas were situated close to each other and were slightly overlapping in their most medial aspect (Fig. 17B). The space inbetween the three SSEA-1-positive areas appeared to lack the antigen. The positive areas extended from the medially situated dorsal raphe towards the lateral funiculus. They seemed to correspond to Rexed laminae II, IV, and VI/VII (Rexed 1954). It must be noted, however, that the laminae of Rexed are difficult to distinguish in cresyl voilet-stained sections of embryonal spinal cords. In this particular case, the most ventral of the three SSEA-1-positive regions was especially difficult to match with a particular lamina of Rexed. However, at the thoracic level the most ventral positive area included the lateral horn, which suggests that at least part of lamina VII is involved. Between E20 and P2, the SSEA-1 immunoreactivity gradually diminished from the lower two regions. At E20, staining was found to be absent in the most ventral region and around P2, the antigen also appeared to have vanished in the intermediate region. Finally, SSEA-1 continued to be expressed in the most dorsal region. This area was examined more closely in the cresyl violet-stained sections of the postnatal spinal cord and found to correspond to Rexed lamina II. Additionally, during the first few weeks after birth, faint, coarse SSEA-1 immunoreactivity developed throughout the dorsal horn. In the adult rat spinal cord, the overall faint, coarse staining pattern and the intensely stained lamina II remained visible in the dorsal horn (Fig. 17C).

From E16 on, the dorsal raphe stained positively for SSEA-1 (Fig. 17A). The staining intensity was greatest during the embryonal and initial postnatal period. In the adult spinal cord, SSEA-1 could still be detected in the dorsal raphe.

4.3.3.5 Acetylcholinesterase

Before E17, AChE could not be detected in the developing dorsal horn (Fig. 17D). At E18, faint AChE expression appeared to be present in the dorsal horn of the rat spinal cord (Fig. 17E). AChE-positive fibers entered the dorsal horn at its dorsmedial aspect and coursed towards the intermediate region of the mantle layer. In addition, AChE-positive fibers were found to traverse the dorsal raphe towards the contralateral dorsal horn. During the following days of development, the AChE pattern remained essentially similar. After birth, the staining was found to diminish in the decussating fibers described above, and from P8 on they could hardly be recognized. After P8, AChE was pro-

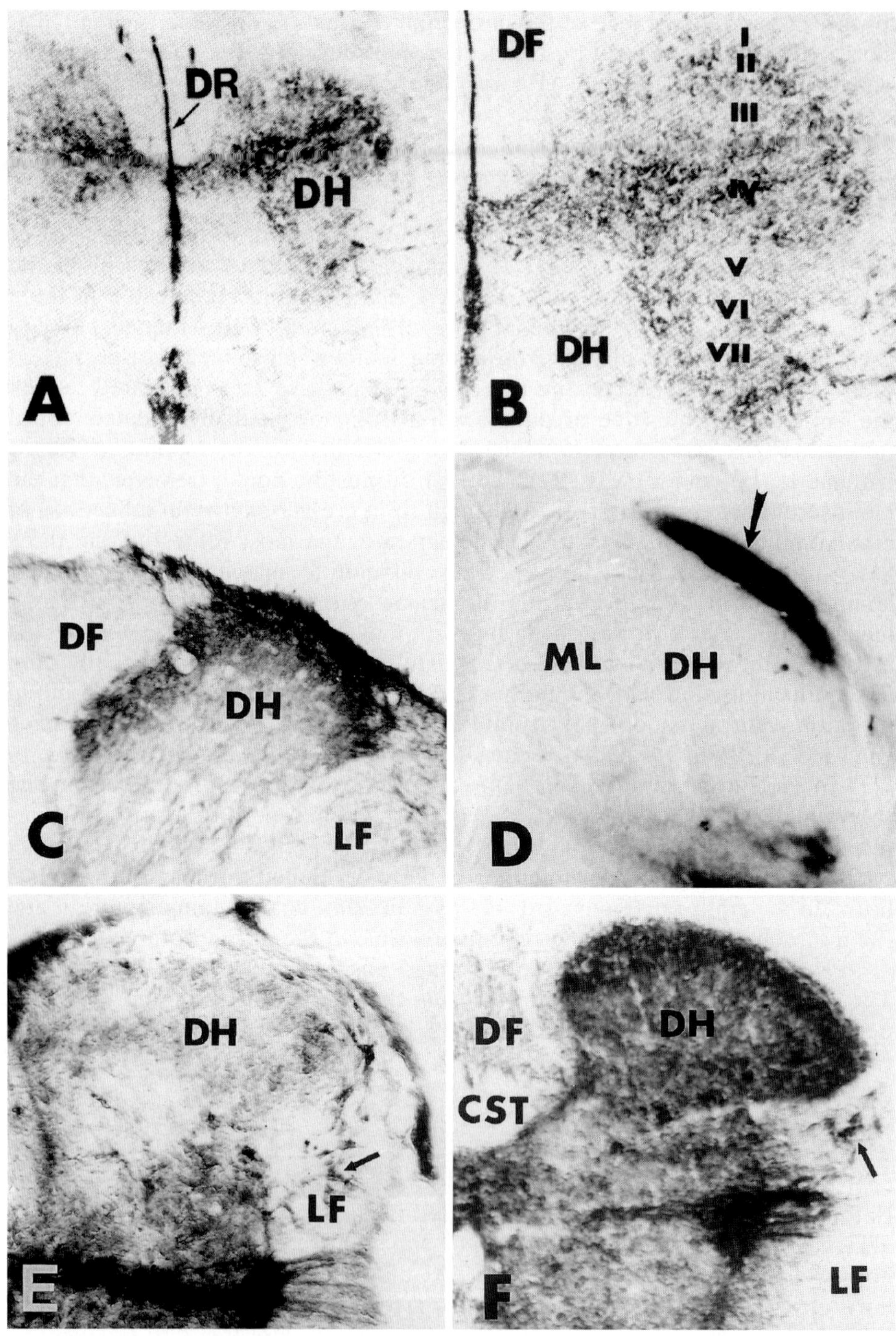

DR
DH
DF
I
II
III
IV
V
VI
VII
DH
A
B
DF
DH
LF
C
ML
DH
D
DH
LF
E
DF
CST
DH
LF
F

minently expressed in the superficial region of the dorsal horn (Fig. 17F). This superficial area was identified in adjacent cresyl violet-stained sections as Rexed's laminae I and II (Rexed 1954). Additionally, after P12, AChE-positive cells were detected scattered throughout the dorsal horn.

From E16 on, intense AChE staining was detected in the dorsal raphe. After birth, the intensity of the staining gradually decreased, and around P20 hardly any enzyme activity could be detected in the dorsal raphe.

4.3.3.6 Vimentin

At E14, a radial pattern of vimentin-positive fibers was found to be present in the dorsal horn (Fig. 18A). These fibers seemed to originate in the dorsal and intermediate part of the matrix layer (see Sect. 4.2.2.6) and were detected in the dorsal horn between E16 and E22. However, after E18, the superficial regions appeared to be less intensely stained (Fig. 18B). During the same period, the staining intensity substantially decreased towards a faint appearance. During the first week after birth, the overall staining intensity further decreased, and from P10 on vimentin immunoreactivity could hardly be detected in the dorsal horn. At this time, some vimentin reactivity was still found scattered in the dorsal horn. This immunoreactivity was found to be present in cells, although single short, immunoreactive protrusions were also detected. At P20, the adult vimentin distribution pattern appeared to be present, with only a single vimentin immunoreactive protrusion in the dorsal horn of the rat spinal cord (Fig. 18C).

At E16, the spinal cord dorsal raphe was found to be intensely stained for the vimentin antigen. During the following few days, however, the intensity of the staining rapidly decreased (Fig. 18B). Nevertheless, in the adult spinal cord, some thin, vimentin-positive fibers could still be detected in the dorsal raphe (Fig. 18C).

4.3.3.7 Glial Fibrillary Acidic Protein

Until birth, apart from some dispersed tangles, the dorsal horn was found to contain only a single GFAP-immunoreactive fiber. After P0, however, more GFAP-positive fibers were detected and, surprisingly, these fibers coursed in a vertical direction, converging towards the neck region of the horn. This is in contrast to the general direction of the GFAP-positive fibers in the other parts

---

Fig. 17A–F. Localization of stage-specific embryonic antigen-1 (**A–C**) and acetylcholinesterase (**D–F**) in the developing dorsal horn (*DH*) of the rat spinal cord. Embryonal day 18 (**A**; E18; *arrow* points at the dorsal raphe, *DR*), E18 (**B**; detail of A; *I–VII* indicate Rexed's lamina), adult (**C**), E15 (**D**; *arrow* points at the dorsal root entrance zone), E18 (**E**; *arrow* points at the positive cells in the dorsolateral funiculus), and postnatal day 16 (**F**; *arrow* points at the positive cells in the dorsolateral funiculus). *DF*, dorsal funiculus; *LF*, lateral funiculus; *ML*, mantle layer; *CST*, corticospinal tract

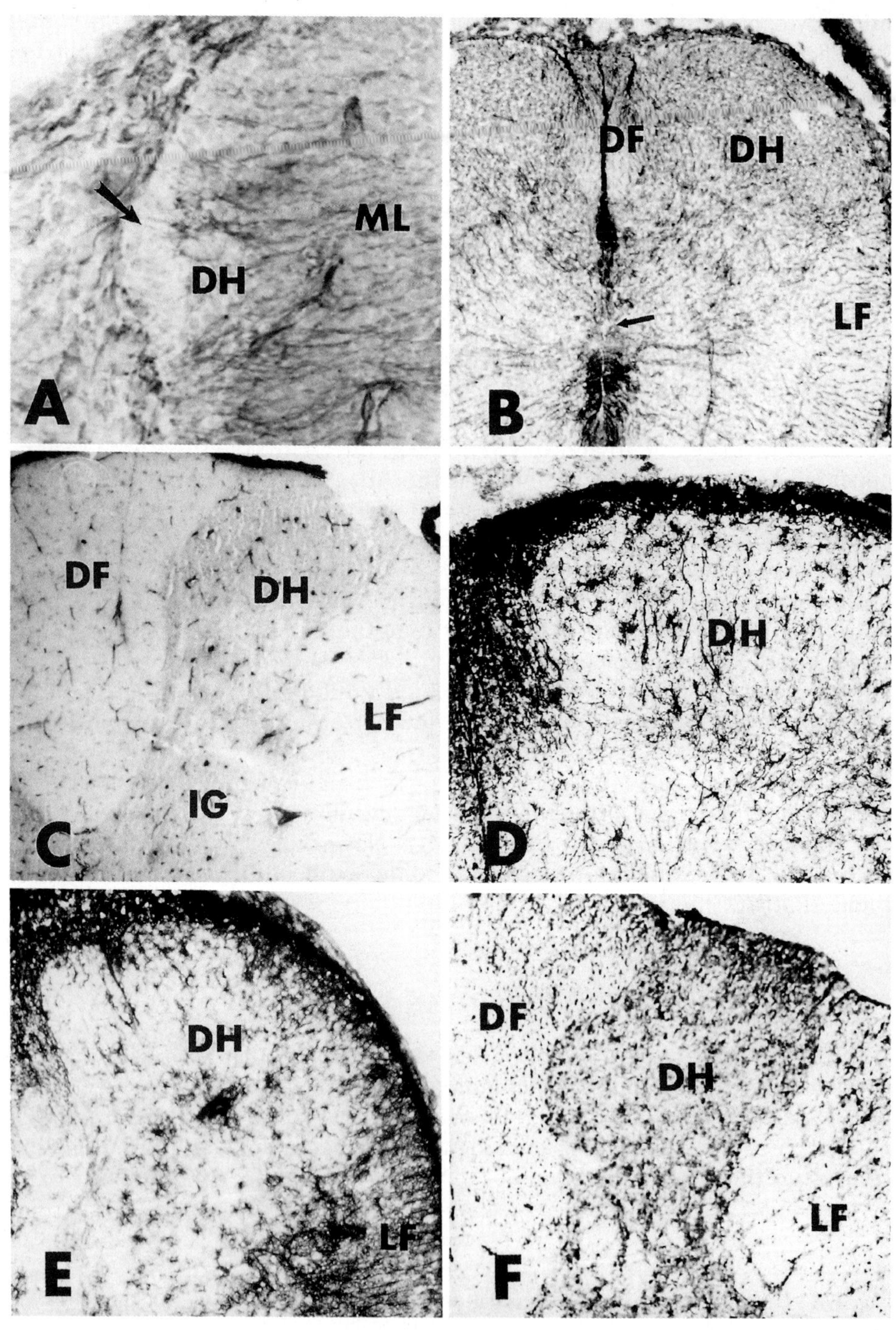

ML
DH
A
DF
DH
LF
B
DF
DH
LF
IG
C
DH
D
DH
LF
E
DF
DH
LF
F

of the mantle layer around this period of development (see Sects. 4.3.1.7 and 4.3.2.8). Between P4 and P10, GFAP staining increased dramatically in the dorsal horn (Fig. 18D,E). Throughout the dorsal mantle layer, short positive protrusions were detected, and in addition an increase in the tangentially orientated GFAP-positive fibers was discerned. GFAP-positive tangles also appeared to be visible in the dorsal horn. At P20, the adult GFAP staining pattern was found to be present, with many positive cells with several short, positive protrusions scattered throughout the dorsal horn (Fig. 18F). Most likely, this distinct GFAP immunoreactivity pattern reflected the presence of the stellate astrocytes of the dorsal horn of the adult spinal cord.

During the embryonal period, GFAP immunoreactivity appeared to be absent in the dorsal raphe. After birth, however, some thin GFAP-positive fibers were found in the dorsal midline (Fig. 18D). These fibers seemed to originate in the intermediate part of the matrix layer (see Sect. 4.2.2.7). In the adult rat spinal cord, GFAP-positive fibers could still be detected in the dorsal raphe.

4.4 Marginal Layer

4.4.1 Ventral Funiculus

4.4.1.1 Neurofilaments

At E13, the first homogeneously distributed, moderately stained neurofilament-positive fibers were detected in the ventral funiculus. The longitudinal fibers were present at all levels of the spinal cord, although they were less abundant in the thoracic and lumbar levels. At E14, a peripheral concentration of neurofilaments was detected in the cervical ventral funiculus (Fig. 19A). Two days later, all levels of the cord displayed more intense neurofilament staining in the ventral funiculus. The staining appeared to be most intense in the periphery. Between E16 and P12, this pattern of neurofilament immunoreactivity remained essentially similar within the ventral funiculus (Fig. 19B). After P12, the staining intensity was found to be lower in the most peripheral region of the ventrolateral part of the cervical ventral funiculus than in its deeper zone (Fig. 19C). Both the superficial neurofilament-positive area and the more heavily stained deeper zone continued from the ventrolateral into the lateral funiculus (see Sect. 4.4.2.1). At maturity, this neurofilament-distribution pattern was observed in the ventral funiculus at all levels of the spinal cord.

Fig. 18A–F. Immunocytochemical localization of vimentin and glial fibrillary acidic protein in the developing dorsal horn (*DH*) of the rat spinal cord. Vimentin at embryonal day 14 (**A**; E14; *arrow* points at the dorsal root entrance zone), E20 (**B**; *arrow* points at the central canal), and postnatal day 20 (**C**; P20). Glial fibrillary acidic protein at P4 (**D**), P12 (**E**), and P20 (**F**). *ML*, matrix layer; *DF*, dorsal funiculus; *LF*, lateral funiculus; *IG*, intermediate gray

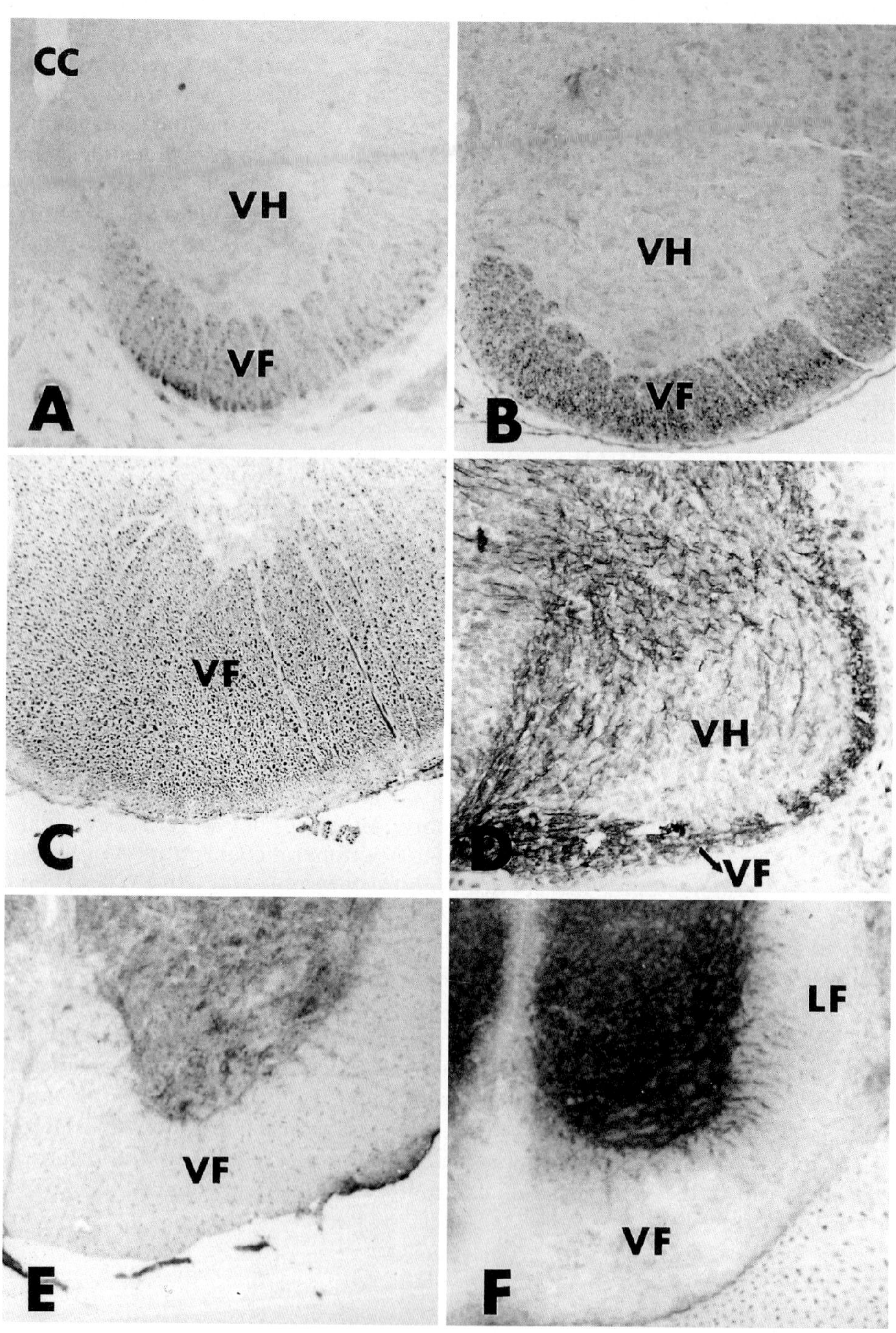
CC
VH
VF
A
VH
VF
B
VF
C
VH
VF
D
VF
E
LF
VF
F

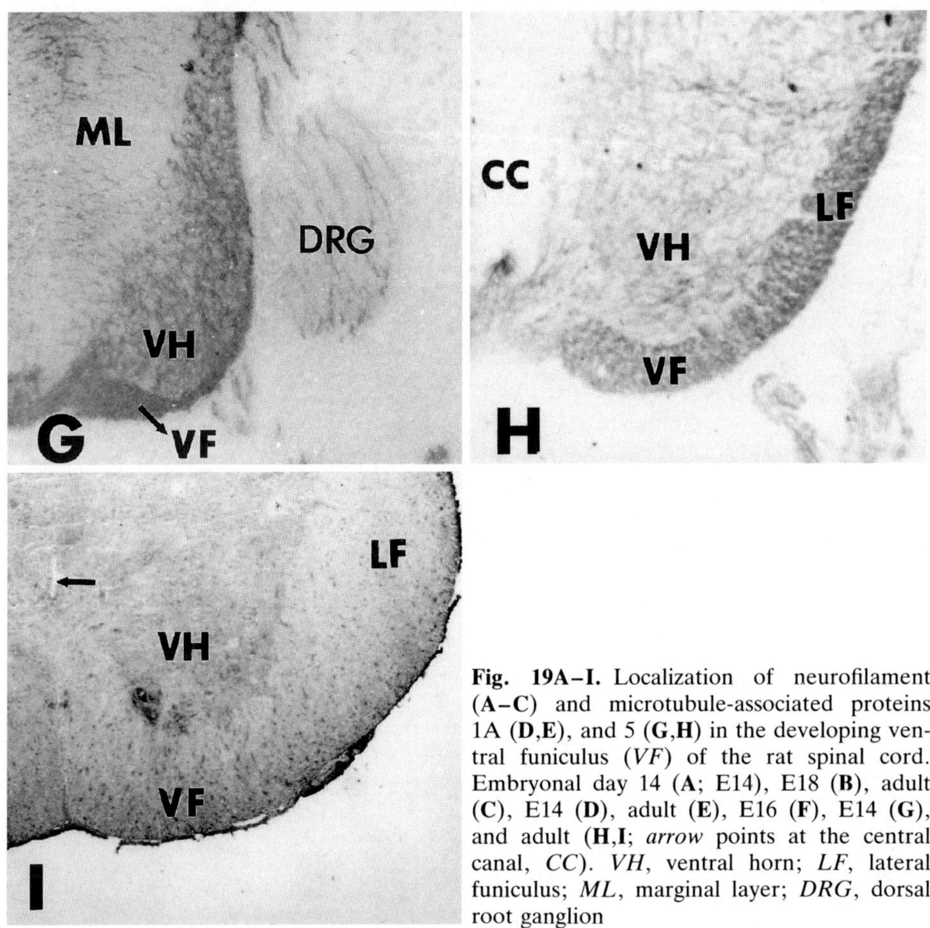

Fig. 19A–I. Localization of neurofilament (A–C) and microtubule-associated proteins 1A (**D,E**), and 5 (**G,H**) in the developing ventral funiculus (*VF*) of the rat spinal cord. Embryonal day 14 (**A**; E14), E18 (**B**), adult (**C**), E14 (**D**), adult (**E**), E16 (**F**), E14 (**G**), and adult (**H,I**; *arrow* points at the central canal, *CC*). *VH*, ventral horn; *LF*, lateral funiculus; *ML*, marginal layer; *DRG*, dorsal root ganglion

4.4.1.2 Microtubule-Associated Proteins

At E14, the longitudinal fibers of the ventral funiculus were found to abundantly express the MAP 1A antigen (Fig. 19D). During the next 2 days of development, the overall homogeneous staining pattern and intensity remained essential similar. From E17 until P10, the staining intensity gradually decreased towards a moderate appearance. From P20 on, faint MAP 1A immunoreactivity could still be discerned in the ventral funiculus. Surprisingly, after P20 small round cells containing MAP 1A could also be detected in the ventral funiculus. The adult ventral funiculus stained faintly for MAP 1A and the described small positive cells were still present (Fig. 19E).

At E16, moderately stained MAP 2-positive processes were found in the ventral funiculus. These processes appeared to originate from cells in the ventral mantle layer and coursed into the ventral funiculus (Fig. 19F). Between E16 and P10, such a MAP 2 staining pattern remained detectable in the ventral

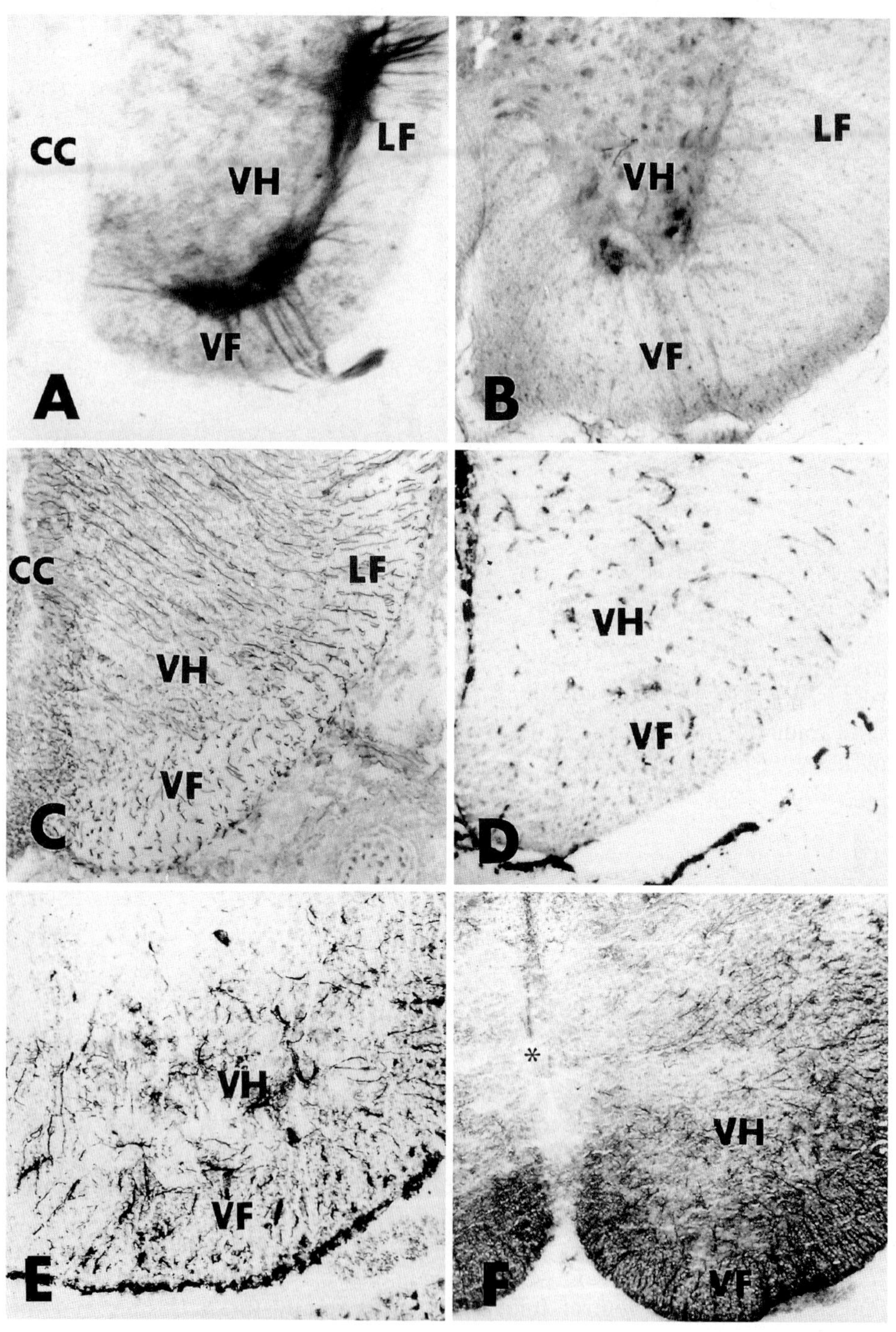
CC
LF
VH
VF
A
LF
VH
VF
B
CC
LF
VH
VF
C
VH
VF
D
VH
VF
E
*
VH
VF
F

funiculus. Thereafter, the staining rapidly vanished, and after P20 hardly any MAP 2 immunoreactivity could be detected in the ventral funiculus (Fig. 19I).

From E13 on, the longitudinal fibers of the ventral funiculus were found to be intensely stained for the MAP 5 antigen (Fig. 19G). The homogeneously distributed MAP 5 immunoreactivity remained visible until P10; thereafter, the staining intensity rapidly declined towards a faint appearance for the MAP 5 antigen at maturity (Fig. 19H).

4.4.1.3 Stage-Specific Embryonic Antigen-1

During the embryonal period of spinal cord development, the ventral funiculus was devoid of SSEA-1 immunoreactivity. After birth, however, faint, coarse SSEA-1 staining could be detected and this persisted in the adult rat spinal cord ventral funiculus.

4.4.1.4 Acetylcholinesterase

At E18, AChE-positive fibers could clearly be discerned in the ventral funiculus (Fig. 20A). These fibers originated from the AChE-positive motor cells of the ventral horn and were found to traverse the funiculus into the periphery. Between E18 and P10, these fibers remained AChE positive, but thereafter they gradually lost their staining activity. From P12 until maturity, a "network" of short, AChE-positive processes was found to be present in the ventral funiculus (Fig. 20B). Most likely, these fibers represented the dendrites of the cells of the ventral and intermediate mantle layer, which appear to extend into the ventral funiculus.

4.4.1.5 Vimentin

At E13, disarranged short vimentin-positive fibers were detected in the ventral funiculus. During the next 2 days of spinal cord development, a clear radial pattern of vimentin-positive fibers was found to develop. Nevertheless, some short disarranged vimentin-positive fibers were still detected. From E15 on, the funiculus filled with radially orientated, intensely stained, vimentin-positive fibers (Fig. 20C). Between E18 and P6, this radial staining pattern remained essentially similar. In general, the intensity of the vimentin staining gradually decreased during this period towards a faint appearance. Around this age, the palisade of vimentin fibers seemed to "withdraw" from the ventral funiculus.

◀

Fig. 20A–F. Expression patterns of acetylcholinesterase (**A,B**), vimentin (**C,D**), glial fibrillary acidic protein (**E,F**) in the developing ventral funiculus (*VF*) of the rat spinal cord. Embryonal day 18 (**A**; E18), adult (**B**), E17 (**C**), adult (**D**), E18 (**E**), and adult (**F**; *star* indicates central canal, *CC*). *VH*, ventral horn; *LF*, lateral funiculus

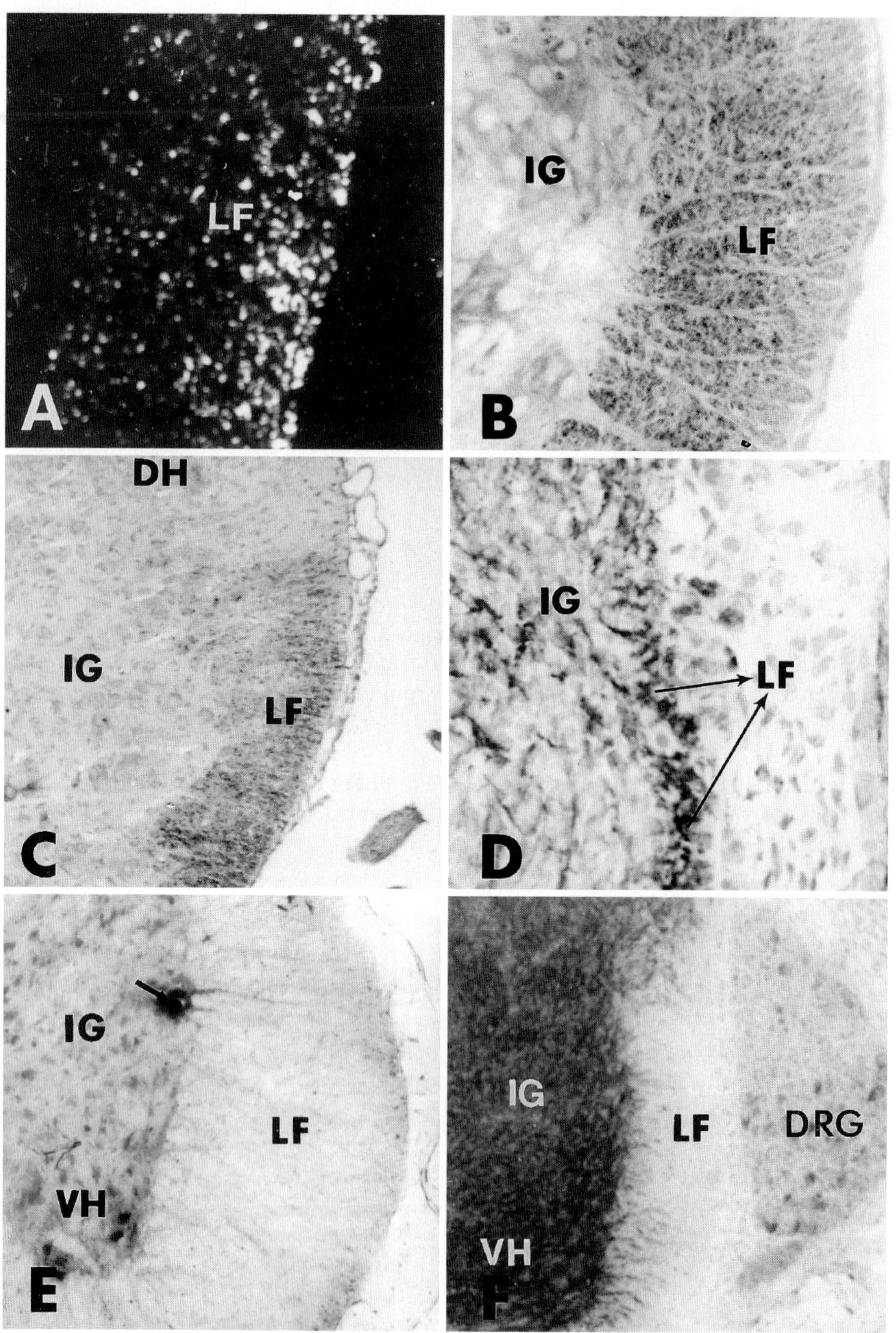

A
LF
B
IG
LF
DH
IG
LF
C
IG
LF
D
E
IG
LF
VH
F
IG
LF
DRG
VH

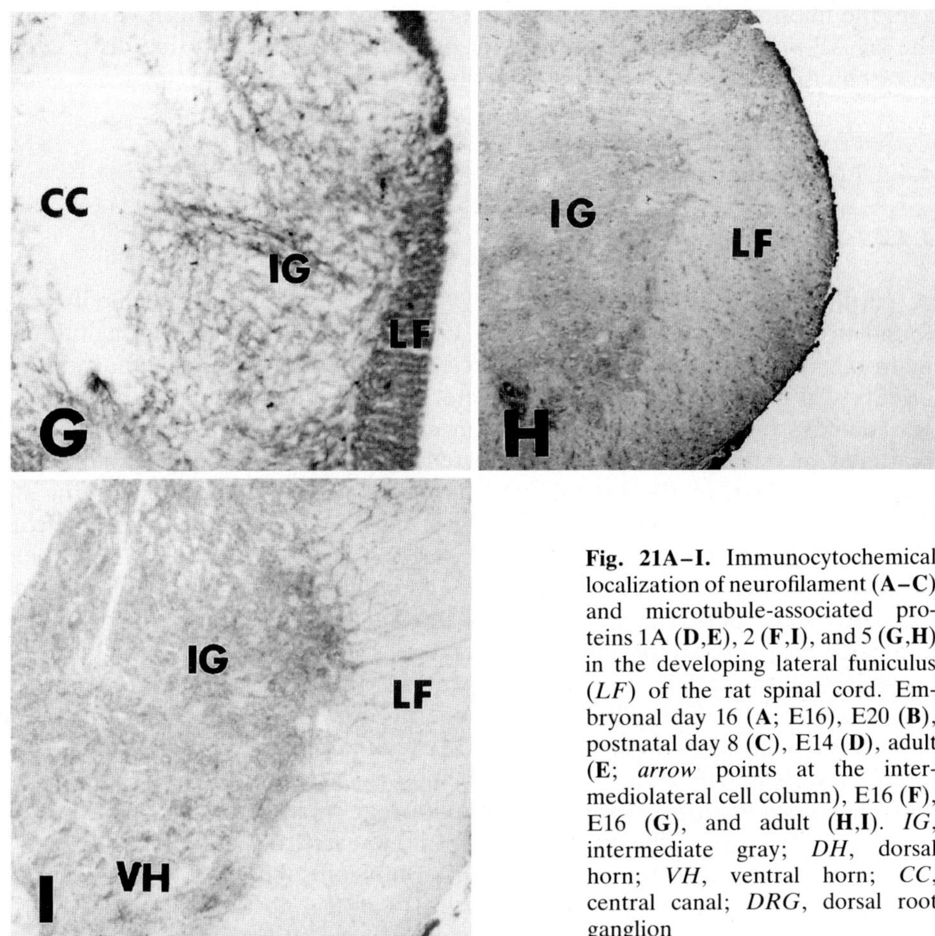

Fig. 21A–I. Immunocytochemical localization of neurofilament (**A–C**) and microtubule-associated proteins 1A (**D,E**), 2 (**F,I**), and 5 (**G,H**) in the developing lateral funiculus (*LF*) of the rat spinal cord. Embryonal day 16 (**A**; E16), E20 (**B**), postnatal day 8 (**C**), E14 (**D**), adult (**E**; *arrow* points at the intermediolateral cell column), E16 (**F**), E16 (**G**), and adult (**H,I**). *IG*, intermediate gray; *DH*, dorsal horn; *VH*, ventral horn; *CC*, central canal; *DRG*, dorsal root ganglion

The staining first seem to disappear in the most peripheral part of the radially orientated fibers and later on in their medial part. At P20, the adult configuration was found to be present, showing a ventral funiculus devoid of vimentin immunoreactivity (Fig. 20D).

4.4.1.6 Glial Fibrillary Acidic Protein

At E18, GFAP-positive, disarranged, short protrusions were found to be present in the ventral funiculus (Fig. 20E). These protrusions originated from the ventrolateral aspect of the ventral horn. In addition, single radially orientated fibers could be detected. At E19, more radial GFAP-positive fibers appeared to be present, and at E21 the ventral funiculus was filled with a palisade of GFAP-positive fibers. Around P6, the ventral funiculus demonstrated an explosive increase in its GFAP staining intensity (Fig. 20E). At this age, the fibers were still arranged in the described palisade pattern. After P20, the GFAP-positive fibers in the ventral funiculus were found to be distributed in essentially the same pattern as described for P6.

4.4.2 Lateral Funiculus

4.4.2.1 Neurofilaments

A peculiar spatiotemporal pattern in the expression of neurofilaments was found to be present in the developing lateral funiculus. At E12, longitudinal neurofilament-positive fibers were detected scattered in the lateral funiculus at all levels of the spinal cord. One day later, uniform neurofilament distribution was demonstrated, except for in the most dorsal part, in which only a few scattered positive fibers could be detected. Between E16 and P12, neurofilament distribution in the lateral funiculus followed an intriguing pattern. At E16, neurofilament-immunoreactive fibers were found to be concentrated in the periphery of the lateral funiculus, especially at the cervical and thoracic levels (Fig. 21A). This pattern changed, and between E18 and P4 more neurofilaments were detected in the region just under the surface of the lateral funiculus (Fig. 21B). The most peripheral region still demonstrated neurofilament immunoreactivity, but clearly less intense than in the deeper zone. After P4, at all levels of the cord, it was the superficial layer again that demonstrated intensified neurofilament staining (Fig. 21C). Finally, from P12 until maturity, more neurofilaments were found in the subsurface region of the lateral funiculus. This positive area formed a band-like region close to the periphery. At all spinal cord levels, the subsurface region appeared to be continuous with a more intensely stained region in the ventral funiculus (see Sect. 4.4.1.1).

A small dorsolaterally located area in the dorsal part of the lateral funiculus was found to be filled with neurofilament-positive fibers around E20. The ventromedial part of the area just beneath the dorsal horn filled with positive fibers around P8. After P8, the dorsal part of the lateral funiculus demonstrated a uniform neurofilament distribution pattern. In the adult rat spinal cord, this region was found to be intensely stained for neurofilaments and was continuous with the intensely stained region in the lateral funiculus (see above).

4.4.2.2 Microtubule-Associated Proteins

At E14, the longitudinal fibers in the lateral funiculus were found to demonstrate abundant MAP 1A staining (Fig. 21D). During the following few days of spinal cord development, the overall homogeneous staining pattern of MAP 1A remained essentially similar. Between E17 and P10, however, the staining intensity gradually declined towards a faint appearance of the antigen. Similar, faint MAP 1A staining was also detected in the mature lateral funiculus (Fig. 21E). After P10, the funiculus was found to contain small round cells, most likely glial cells, which clearly stained positively for the MAP 1A antigen.

At E16, intensely stained MAP 2-positive processes were found to pro-
trude into the lateral funiculus (Fig. 21F). These fibers seemed to originate in
the intermediate part of the mantle layer. After E18, the most dorsal aspect of
the lateral funiculus in particular, just beneath the dorsal horn, was found to
contain a large number of such MAP 2-positive processes. Between E18 and
P10, the MAP 2 distribution pattern remained essentially similar. After P10,
however, the staining was found to gradually disappear, and at maturity the
lateral funiculus was almost devoid of MAP 2 immunoreactivity (Fig. 21I).

From E13 on, the lateral funiculus was found to contain longitudinal MAP
5-positive fibers. The intense MAP 5 immunoreactivity was homogeneously
distributed throughout the funiculus. Between E13 and P10, the distribution
pattern remained essentially similar, but thereafter its intensity gradually
declined (Fig. 21G). Around P20, a faint homogeneous MAP 5 distribution
could still be demonstrated in the lateral funiculus. This faint appearance was
also found to be present in the adult lateral funiculus (Fig. 21H).

4.4.2.3 Stage-Specific Embryonic Antigen-1

At E13–E14, SSEA-1 was detected in a small region dorsally located in the
lateral funiculus (Fig. 22A). One day later, this region appeared to lack SSEA-
1 staining. During the rest of the embryonal period, SSEA-1 immunoreactivity
was found to be absent from the lateral funiculus. After birth, however, the
lateral funiculus developed a faint, coarse SSEA-1 staining pattern. Similar,
coarse staining was also detected in the adult spinal cord lateral funiculus.

4.4.2.4 Acetylcholinesterase

At E18, AChE-positive fibers were found to be present in the lateral funiculus.
These fibers originated from AChE-positive cells located in the most lateral
part of the intermediate mantle layer (Fig. 22B). Besides these fibers, heavily
stained AChE-positive cells were also detected in the lateral funiculus, located
just beneath the dorsal horn (Fig. 22B; also see Sect. 4.3.2.6). During the following days of embryonal development and during the first 2 weeks after
birth, the described AChE pattern remained essentially similar. From P12
on, in addition to the clearly recognizable fibers from the laterally located
cells in the intermediate gray, short AChE-positive fibers appeared to be
present in the lateral funiculus (Fig. 22C). These short fibers, most likely
dendrites from the cells of the ventral and intermediate region of the mantle
layer, were found to form an extensive "network" in the lateral funiculus. This
distribution pattern of AChE was also detected in the lateral funiculus of the
adult spinal cord.

4.4.2.5 Vimentin

At E13 and E14, short vimentin positive fibers were found scattered throughout the lateral funiculus. At E15, however, this distribution pattern was largely

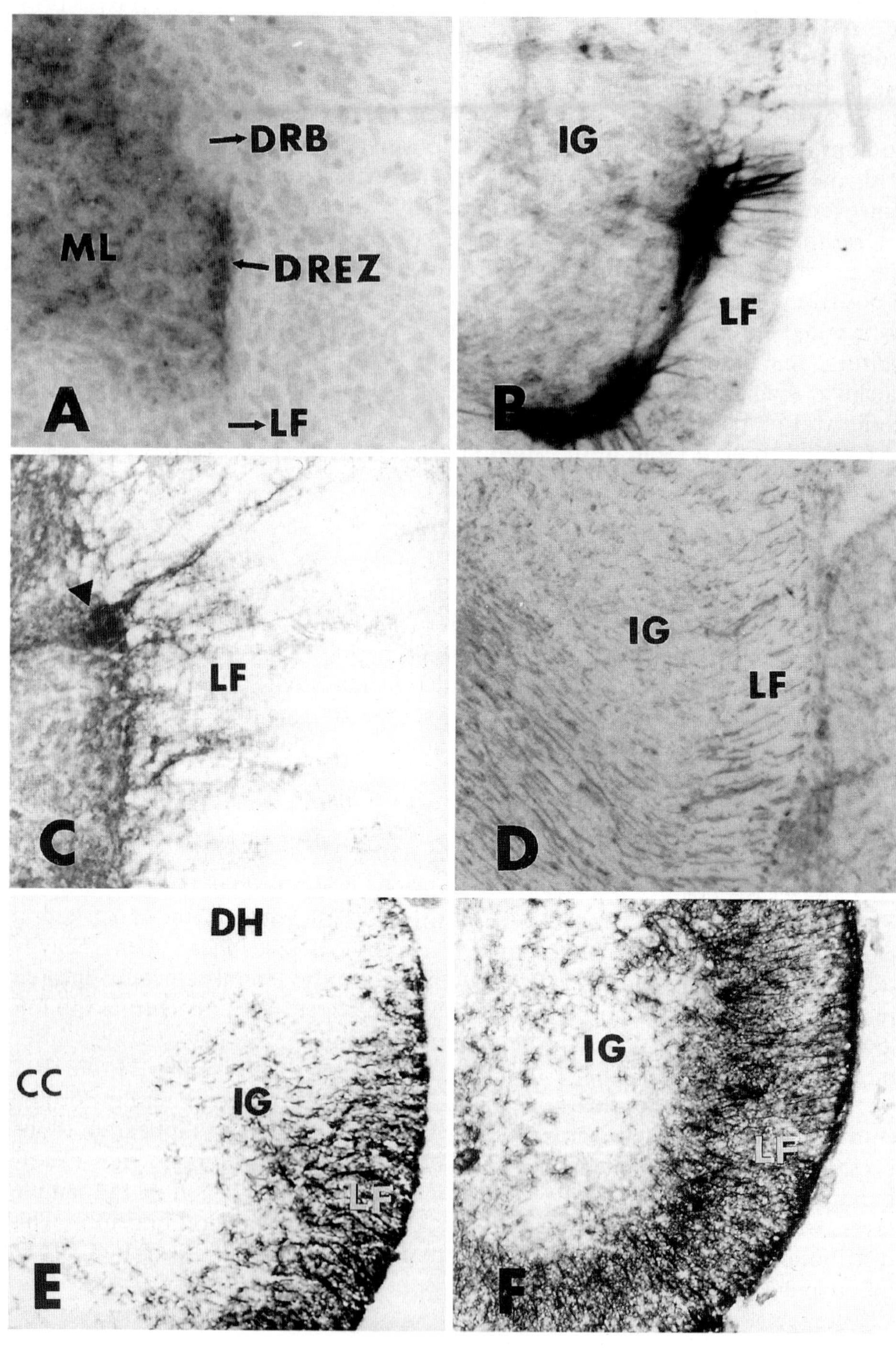

DRB
ML
DREZ
LF
A
IG
LF
B
LF
C
IG
LF
D
DH
CC
IG
LF
E
IG
LF
F

replaced by a clear palisade of vimentin-positive fibers. Nevertheless, some short, disarranged, vimentin-positive protrusions were still observed. Until P6, the staining pattern remained essentially similar, but the overall intensity gradually declined (Fig. 22D). After P6, vimentin staining was found to further decrease in the lateral funiculus. The radially orientated fibers seemed to withdraw from the lateral funiculus, i.e., the activity first vanished from their most peripheral part and later on from their medial part. At P20, the adult staining pattern in the lateral funiculus was found to be present, with virtually no vimentin immunoreactivity.

4.4.2.6 Glial Fibrillary Acidic Protein

At E18, short GFAP-positive protrusions were detected scattered throughout the ventral part of the lateral funiculus. These protrusions seemed to originate from the ventrolateral part of the ventral horn. Some radially orientated GFAP-positive fibers were also found to be present. At E21, instead of the disarranged pattern, the lateral funiculus was found to be filled with an arranged palisade of GFAP-positive fibers (Fig. 22E). After P6, an explosive increase in the amount of GFAP-positive fibers occurred in the lateral funiculus (Fig. 22F). This pattern of GFAP-positive fibers was also found in the adult rat spinal cord.

4.4.3 Dorsal Funiculus/Dorsolateral Fasciculus

4.4.3.1 Neurofilaments

In E13 rat embryos, the dorsal root bifurcation zone demonstrated a distinct neurofilament staining pattern (Fig. 23A). The dorsal part and a small ventrally situated area stained more intensely than the medial part. During the following days of spinal cord development, the bifurcation zone and its fibers seemed to shift medially, and at E16 an initial dorsal funiculus could be discerned. At this age, neurofilament-positive fibers were found to be present throughout the dorsal funiculus, with a concentration of them in its most medial part (Fig. 23B). During the following days, the now recognizable fasciculus gracilis contained more neurofilaments in its most dorsal part (Fig. 23C). After P0, however, it demonstrated a homogeneous distribution of neurofilament immunoreactivity. From E18 on, the lateral part of the fasciculus cuneatus exhibited more intense staining, apparently due to the ingrowth of the dorsal root fibers. In the mature spinal cord, neurofilament staining was more pronounced in the entire fasciculus cuneatus than in the fasciculus gracilis, which seemed to

Fig. 22A–F. Developmental expression patterns of stage-specific embryonic antigen-1 (**A**), acetylcholinesterase (**B,C**), vimentin (**D**), and glial fibrillary acidic protein (**E,F**) in the developing lateral funiculus (*LF*) of the rat spinal cord. Embryonal day 14 (A; E14), E18 (**B**), postnatal 20 (C; P20; *arrowhead* points at the intermediolateral cell column), E17 (**D**), E20 (**E**), and P16 (**F**). *ML*, matrix layer; *DRB*, dorsal root bifurcation zone; *DREZ*, dorsal root entrance zone; *IG*, intermediate gray; *DH*, dorsal horn; *CC*, central canal

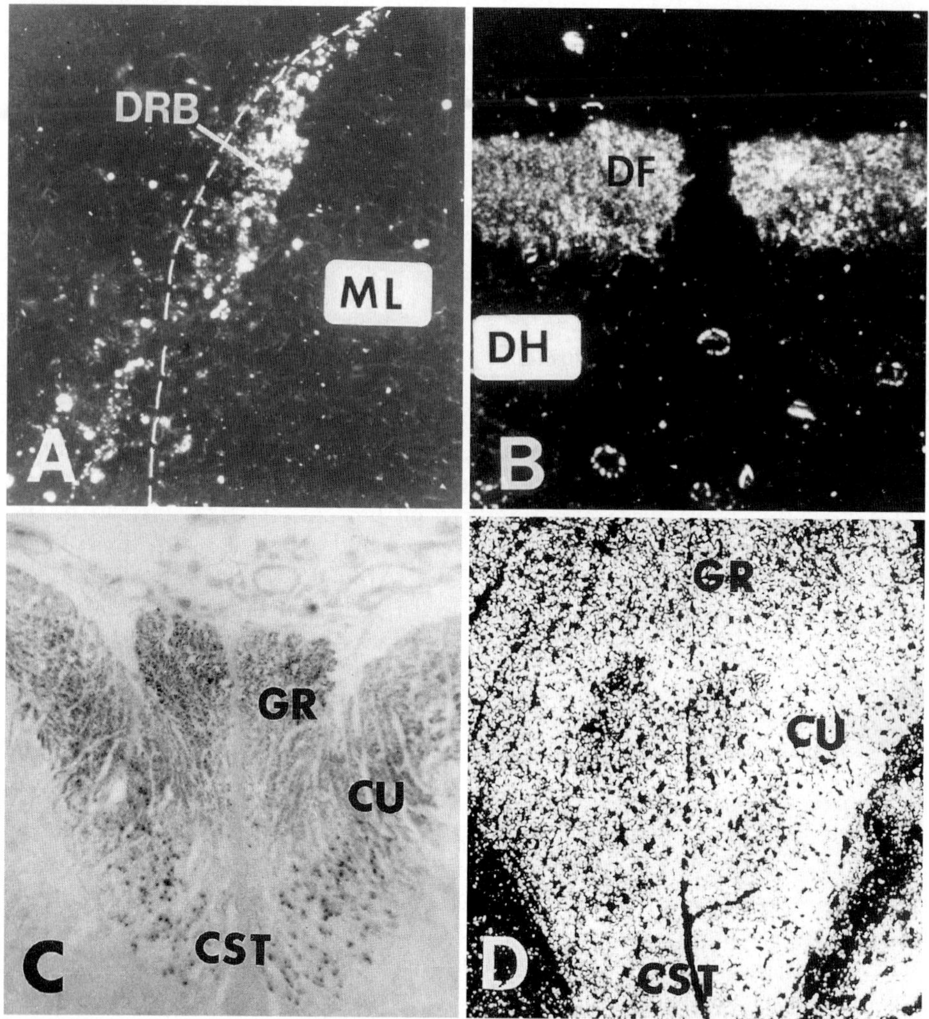

Fig. 23A–D. Neurofilament expression patterns in the developing dorsal funiculus (*DF*) of the rat spinal cord. Dark field microphotographs at embryonal day 13 (**A**; E13), E16 (**B**), and in adult (**D**). **C** Bright field microphotograph of E20. *DRB*, dorsal root bifurcation zone; *ML*, matrix layer; *DH*, dorsal horn; *CST*, corticospinal tract; *CU*, fasciculus cuneatus; *GR*, fasciculus gracilis

be caused by the presence of a higher number of thick fibers in the former region (Fig. 23D).

At E18, the ventral part of the dorsal funiculus demonstrated the first intensely stained longitudinal, neurofilament-positive fibers (Fig. 23C). During further development, the region gradually filled with fibers, according to a rostral-to-caudal gradient. Twelve days after birth, the most ventral part of the dorsal funiculus demonstrated a homogeneous distribution of neurofilament-positive axons at all levels of the spinal cord (Fig. 23D).

At E16, the first longitudinal neurofilament-positive fibers were detected in the dorsolateral fasciculus. Although faintly stained, this small region remained positive for neurofilaments during the following days of development as well as in the adult rat spinal cord.

4.4.3.2 Microtubule-Associated Proteins

At E14, the dorsal root entrance zone and the dorsal root bifurcation zone displayed MAP 1A-positive fibers (Fig. 24A). During the next 2 days of spinal cord development, moderately stained fibers remained present in the developing dorsal funiculus. At E18, the now distinguishable fasciculus cuneatus was found to exhibit MAP 1A-positive fibers, whereas the fibers of the fasciculus gracilis were devoid of this antigen (Fig. 24B). After P10, both the fasciculus gracilis and the fasciculus cuneatus lacked MAP 1A immunostaining. Small MAP 1A-positive cells were detected throughout the dorsal columns (Fig. 24C).

Between E20 and P10, the fibers in the ventral part of the dorsal funiculus gradually developed MAP 1A positivity. From P20 until maturity, MAP 1A remained detectable in the ventral dorsal funiculus (Fig. 24C).

Around E16, MAP 1A-positive fibers were found in the dorsolateral fasciculus. During postnatal development as well as in the adult rat spinal cord, faintly stained MAP 1A-positive fibers remained present in the dorsolateral fasciculus.

During the embryonal and postnatal period of spinal cord development, the dorsal funiculus and the dorsolateral fasciculus were found to be devoid of MAP 2 immunoreactivity (Fig. 24D–F). The most ventral part of the dorsal funiculus also appeared to lack MAP 2 (Fig. 24D–F).

Between E13 and E16, the fibers in the dorsal root entrance zone and the dorsal root bifurcation zone stained intensely for the MAP 5 antigen (Fig. 24G). Similar, intense MAP 5 staining was discerned throughout the dorsal funiculus after the development of its medial fasciculus gracilis and lateral fasciculus cuneatus (Fig. 24H). Between E18 and P10, the fibers in both the fasciculus gracilis and cuneatus remained intensely stained, but after P10 their intensity for the MAP 5 antigen gradually decreased towards only a faint appearance at maturity (Fig. 24I).

From E18 on, the fibers in the most ventral part of the dorsal funiculus were intensely stained for MAP 5 (Fig. 24H). During the postnatal period of spinal cord development and in the mature cord, such an abundant MAP 5 immunostaining remained present in the ventral funiculus (Fig. 24I).

At E16, the dorsolateral fasciculus demonstrated faint MAP 5 immunoreactivity. A similar, faint staining could still be detected in the adult dorsolateral fasciculus.

4.4.3.3 Stage-Specific Embryonic Antigen-1

At E14, the dorsal root entrance zone was found to exhibit moderate SSEA-1 staining. One day later, however, SSEA-1 immunoreactivity could hardly be discerned in this region. At E18, only a positive dorsal raphe was detected in

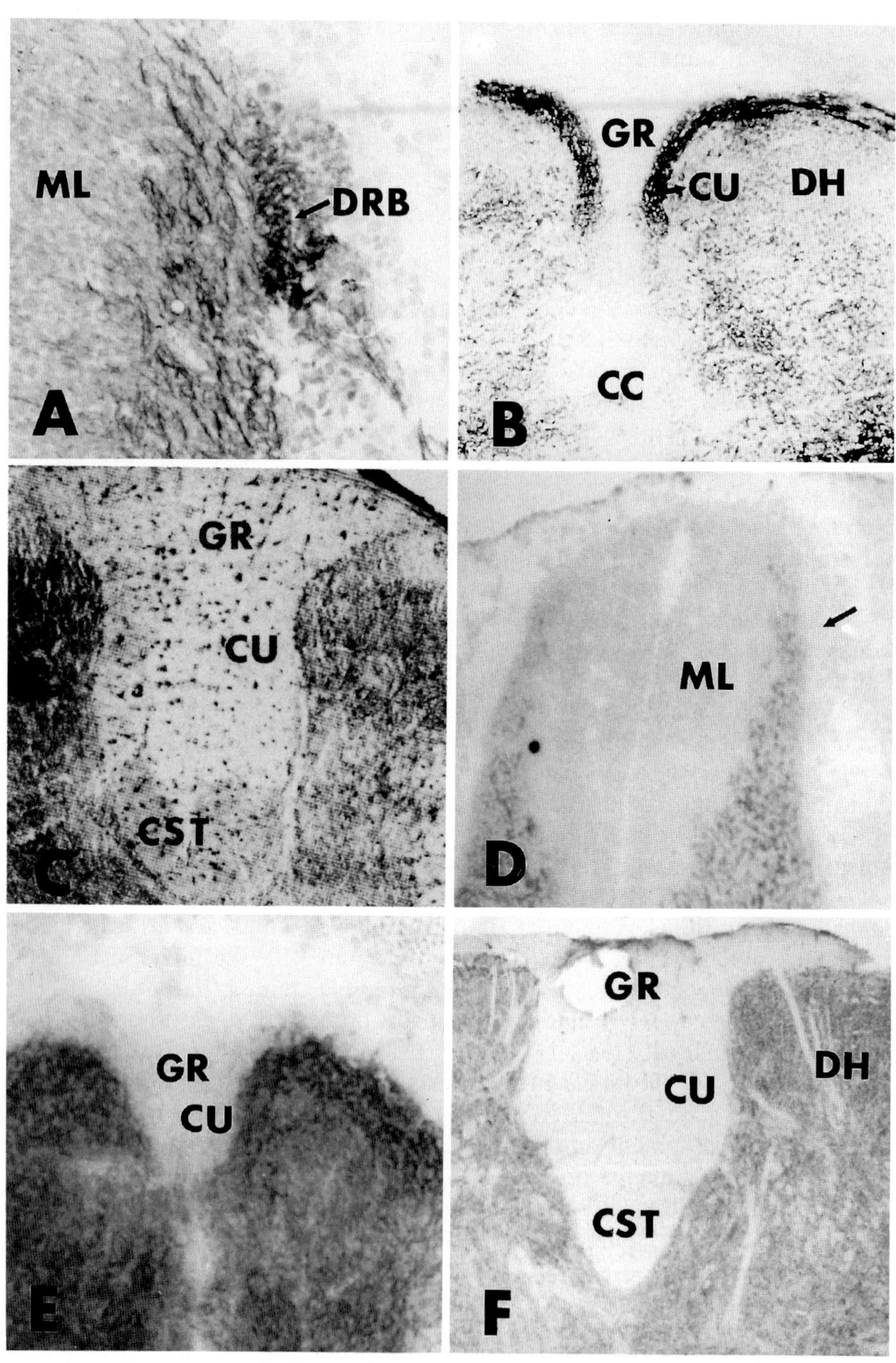

ML
DRB
GR
CU
DH
CC
A
B
GR
CU
CST
ML
C
D
GR
CU
GR
CU
DH
CST
E
F

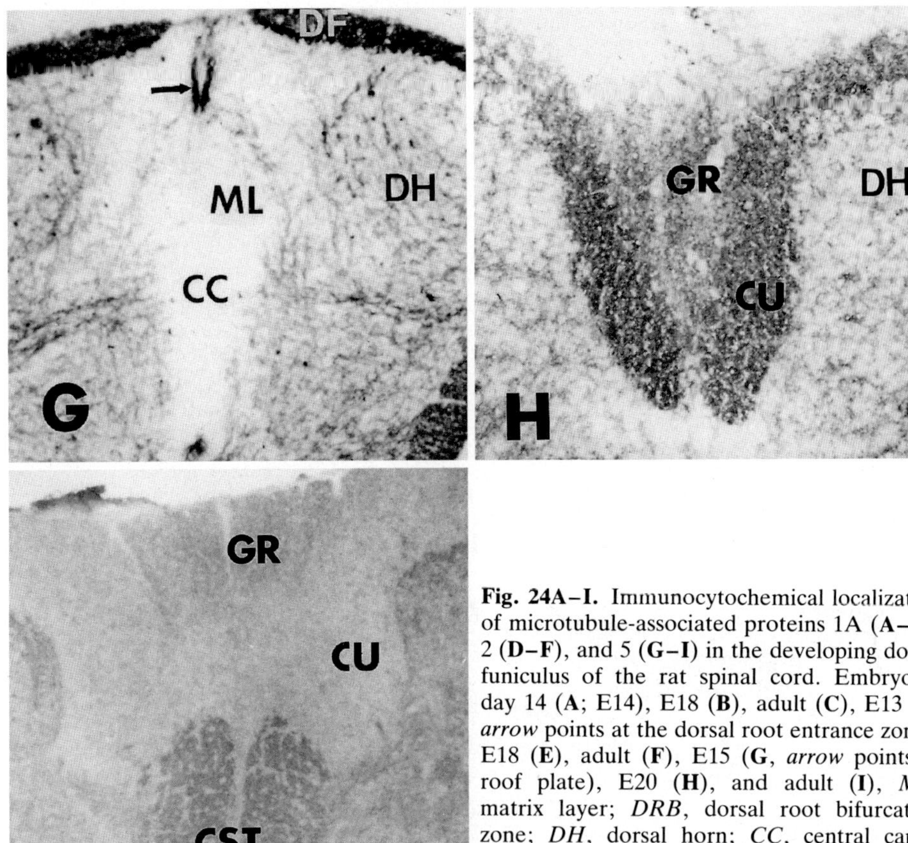

Fig. 24A–I. Immunocytochemical localization of microtubule-associated proteins 1A (**A–C**), 2 (**D–F**), and 5 (**G–I**) in the developing dorsal funiculus of the rat spinal cord. Embryonal day 14 (**A**; E14), E18 (**B**), adult (**C**), E13 (**D**; *arrow* points at the dorsal root entrance zone), E18 (**E**), adult (**F**), E15 (**G**, *arrow* points at roof plate), E20 (**H**), and adult (**I**), *ML*, matrix layer; *DRB*, dorsal root bifurcation zone; *DH*, dorsal horn; *CC*, central canal; *CST*, corticospinal tract; *CU*, fasciculus cuneatus; *GR*, fasciculus gracilis; *DF*, dorsal funiculus

the dorsal funiculus (Fig. 25E). After birth, an overall faint, coarse SSEA-1 immunoreactivity was found in the dorsal funiculus (including its ventral part) as well as in the dorsolateral fasciculus.

4.4.3.4 Acetylcholinesterase

At E13–E14, the dorsal root entrance zone and the dorsal root bifurcation zone were found to stain abundantly for AChE (Fig. 25A). Three days later, at E16, heavy staining persisted in the dorsal funiculus (Fig. 25B). Most likely, this staining was due to the ingrowing ascending collaterals of the primary afferents, as these fibers were also found to express AChE. At E18, the fasciculus gracilis and the fasciculus cuneatus demonstrated the enzyme (Fig. 25C). Surprisingly, a small dorsal region in the fasciculus gracilis was often found to lack AChE. A similar, negative area was most clearly detected at the thoracic level of the spinal cord. During the final embryonal days, AChE

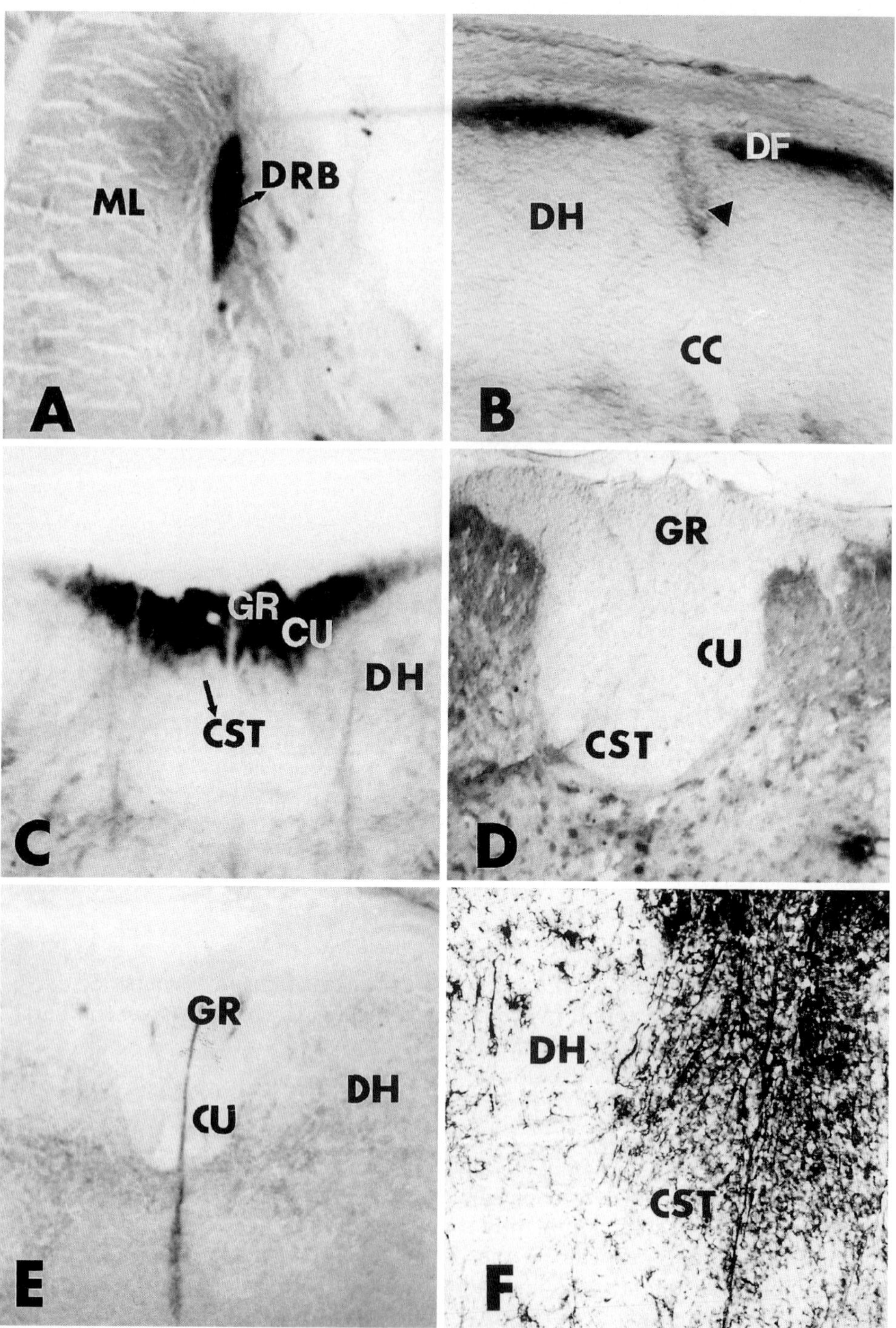
A
ML
DRB
B
DF
DH
CC
C
GR
CU
DH
CST
D
GR
CU
CST
E
GR
CU
DH
F
DH
CST

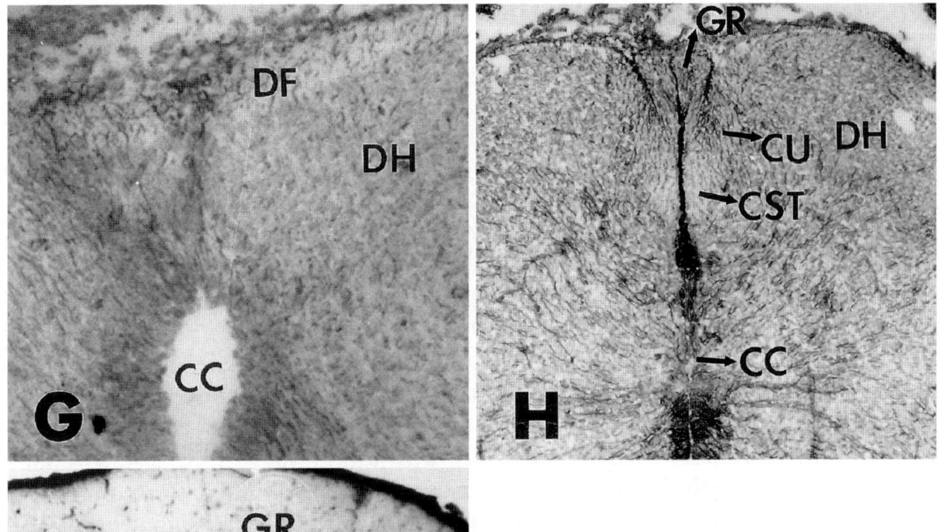
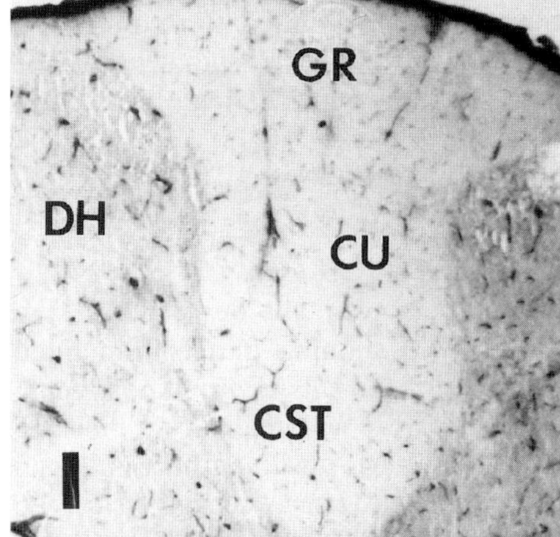

Fig. 25A–I. Expression patterns of acetyl-cholinesterase **(A–D)**, stage-specific embryonic antigen-1 **(E)**, glial fibrillary acidic protein **(F,I)**, and vimentin **(G,H)** in the dorsal funiculus (*DF*) of the developing rat spinal cord. Embryonal day 14 (**A**; E14), E16 (**B**; *arrowhead* points at dorsal raphe), E18 (**C**), adult (**D**), E18 (**E,F**), E16 (**G**), E20 (**H**), and postnatal day 20 (**I**). *ML*, matrix layer; *DRB*, dorsal rootbifurcation zone; *DH*, dorsal horn; *CC*, central canal; *CST*, corticospinal tract; *CU*, fasciculus cuneatus; *GR*, fasciculus gracilis

gradually vanished from the fasciculus gracilis. In general, during the first postnatal week, AChE activity gradually decreased throughout the dorsal funiculus towards a faint appearance. At the end of the second postnatal week, AChE could not be detected in the dorsal funiculus (Fig. 25D).

During the embryonal as well as the postnatal period of spinal cord development, the fibers of the ventral part of the dorsal funiculus were found to lack AChE (Fig. 25C,D).

At E16, the dorsolateral fasciculus was found to demonstrate moderate AChE staining. Between E18 and P8, the staining intensity gradually decreased towards a faint AChE appearance. After P20, the fibers of the dorsolateral fasciculus were devoid of AChE.

4.4.3.5 Vimentin

Between E13 and E15, only a few vimentin-positive fibers were found scattered in the dorsal root entrance zone and bifurcation zone. At E16, however, a palisade of vimentin-positive fibers was discerned (Fig. 25G). After the development of the two fasciculi, the vimentin-positive fibers coursed in a vertical direction towards the ventral part of the funiculus (Fig. 25H). Between E18 and P6, the vimentin staining pattern remained essentially similar, but its intensity gradually decreased towards a faint appearance. Around P6, the vertical orientated fibers clearly converged towards the ventral apex of the dorsal funiculus. At P20 and in the adult spinal cord, only a few vimentin-positive fibers were found scattered in the dorsal funiculus.

At E20, the ventral part of the dorsal funiculus demonstrated some radially orientated vimentin-positive fibers (Fig. 25H). Between E20 and P20, this distribution pattern remained essentially similar, but its intensity substantially decreased. In the adult spinal cord, some radially orientated fibers could still be detected in the ventral part of the dorsal funiculus as well as single short, vimentin-positive protrusions.

At E16, the dorsolateral fasciculus was found to exhibit a few small, disarranged, vimentin-immunoreactive fibers. This staining pattern remained during the following days of development. In the adult rat spinal cord, the dorsolateral fasciculus was found to demonstrate only a single vimentin-immunoreactive fiber.

4.4.3.6 Glial Fibrillary Acidic Protein

After E19, a radial pattern of GFAP-positive fibers gradually developed in the dorsal funiculus (Fig. 25F). Until P6, the GFAP staining pattern remained essentially the same, but thereafter the intensity of the GFAP immunostaining was found to increase dramatically. The fibers were arranged in a palisade pattern. At P20 and in the adult spinal cord, the GFAP-positive fibers in dorsal funiculus appeared to be distributed in the same pattern as described for P6 (Fig. 25I).

Between E18 and P20, as well as short scattered GFAP-positive protrusions, some radially orientated fibers were also detected in the ventral part of the dorsal funiculus. This GFAP distribution pattern was also present in the ventral dorsal funiculus of the adult rat spinal cord.

From P4 on, the dorsolateral fasciculus demonstrated short GFAP-positive protrusions. After P20, however, hardly any GFAP immunoreactivity appeared to be present in this region.

4.5 Tracer Experiments

In all experiments, retrogradely labeled cells were present in the vestibular nuclei, specifically in the lateral vestibular nucleus. Labeling of the medial and descending vestibular nuclei lagged a few days behind the labeling of the lateral vestibular nucleus and was initially confined to the caudalmost parts of the medial and descending nuclei. In all experiments except one (C4258.2; see Table 3), the raphe nuclei were retrogradely labeled. The nucleus raphe magnus was labeled in all experiments, and the nucleus raphe obscurus and pallidus followed a few days later.

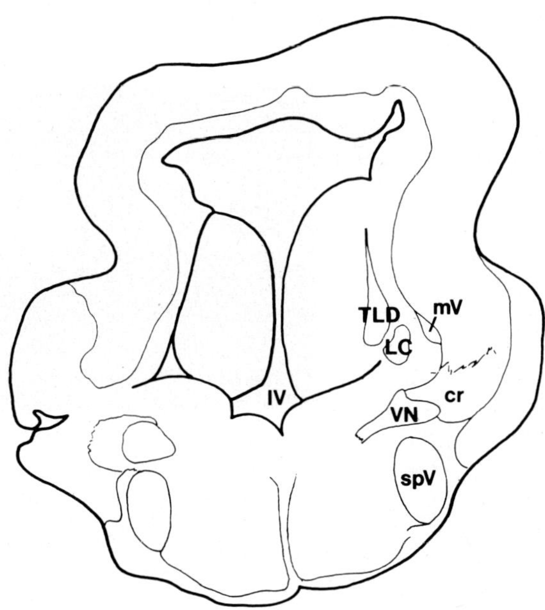

Fig. 26. Transversal sections through the fetal brain result in horizontal-like sections through the pontine region, due to the presence of the pontine flexure. In this section, medulla oblongata, pons, and mesencephalon are all present. The pontine tegmentum, containing the nucleus tegmentalis laterodorsalis (*TLD*), locus coeruleus (*LC*), and mesencephalic trigeminal nucleus (*mV*), bulges into the floor of the fourth ventricular (*IV*) space. In many cases, retrogradely labeled cells in the mV are visible as a string of cells running along the pontine tegmental border from the level of the LC all the way to the top of the mesencephalic ventricle. The LC is located slightly caudally from the TLD and in between TLD and mV. The lateral recess of the fourth ventricle often separates rostral LC from caudal LC. *VN*, vestibular nuclei; *cr*, corpus restiforme; *spV*, spinal trigeminal nucleus

4.5.1 Nucleus Tegmentalis Laterodorsalis

Due to the presence of the mesencephalic and pontine flexures, transverse sections through the pons and rostral mesencephalon of younger fetuses resembled horizontal sections through these regions in the adult rat (Fig. 26). At E17, the TLD was easily distinguished within the pontine central gray (in neutral red-stained sections) as a tear-shaped nucleus, due to the cohesion and arrangement of its constituent neurons. In fact, this tear-shaped nucleus only represented Barrington's subnucleus, and not the whole TLD. In the apparently horizontal sections, the tapered end of the TLD pointed dorsally, towards the mesencephalic periaqueductal gray. The results of the retrograde labeling experiments are summarized in Table 3 (page 24).

At E17, a large injection involving all of the cervical spinal cord as well as the lower medulla oblongata clearly delineated the TLD through retrograde labeling of most of its neurons (C3523). Retrograde labeling of neurons throughout the TLD resulted from injections into the cervical and thoracic spinal cord as well (C4107.3, C3531, and C4258.3; Fig. 27), though the number

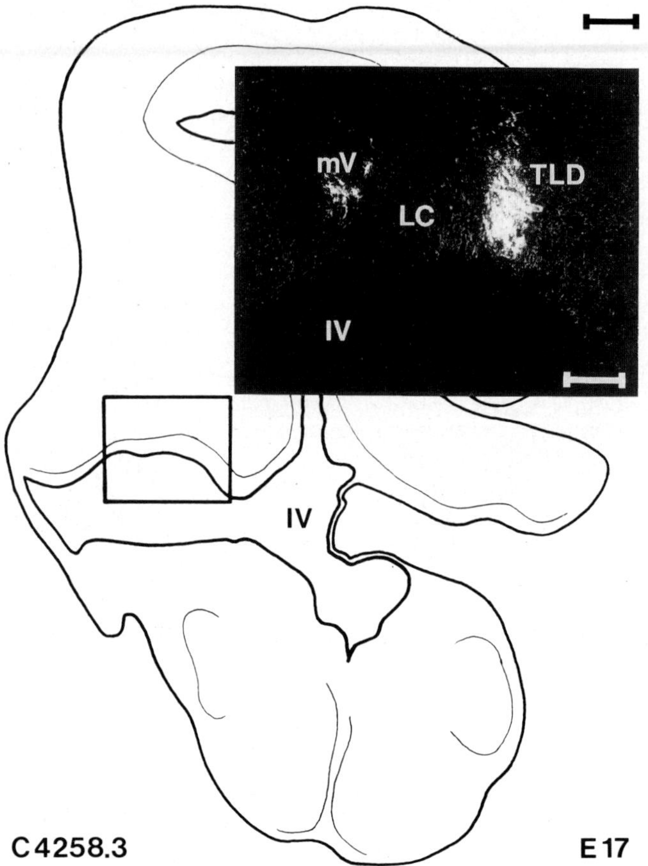

Fig. 27. Early labeling in nucleus tegmentalis laterodorsalis (*TLD*) and mesencephalic trigeminal nucleus (*mV*) resulting from a thoracic injection at embryonal day 16 (E16). No labeling is present in the locus coeruleus (*LC*), located in between the TLD and mV. *Black bar*, 250 μm; *white bar*, 100 μm. *IV*, fourth ventricle

of labeled neurons appeared to be lower. No retrograde labeling was obtained after an injection into the lower lumbar spinal cord (C4258.2).

At E18, injections located cervically caused heavy labeling of neurons throughout the TLD (C3315 and C3316; Fig. 28A). A few scattered retrogradely labeled neurons were observed after an injection located mainly in the cervical cord gray matter (C4162.3). Labeled neurons from injections located in the thoracic and upper lumbar spinal cord appeared progressively less in number, but were always located throughout the dorsoventral extent of the nucleus (C4109.2; Fig. 28B, C4098.5, C4157.1; Fig. 28C, C4162.1). However, after a unilateral injection in the upper lumbar segments, only a few cells were labeled, located ventrally in the ipsilateral TLD (C4169.1). From E19 onwards, all injections into the lumbar spinal cord consistently resulted in retrograde labeling throughout the TLD, though the number of labeled neurons increased with age. At E21, most neurons in the TLD were labeled after an injection into the upper lumbar spinal cord (C4187.5), though an injection mainly confined to

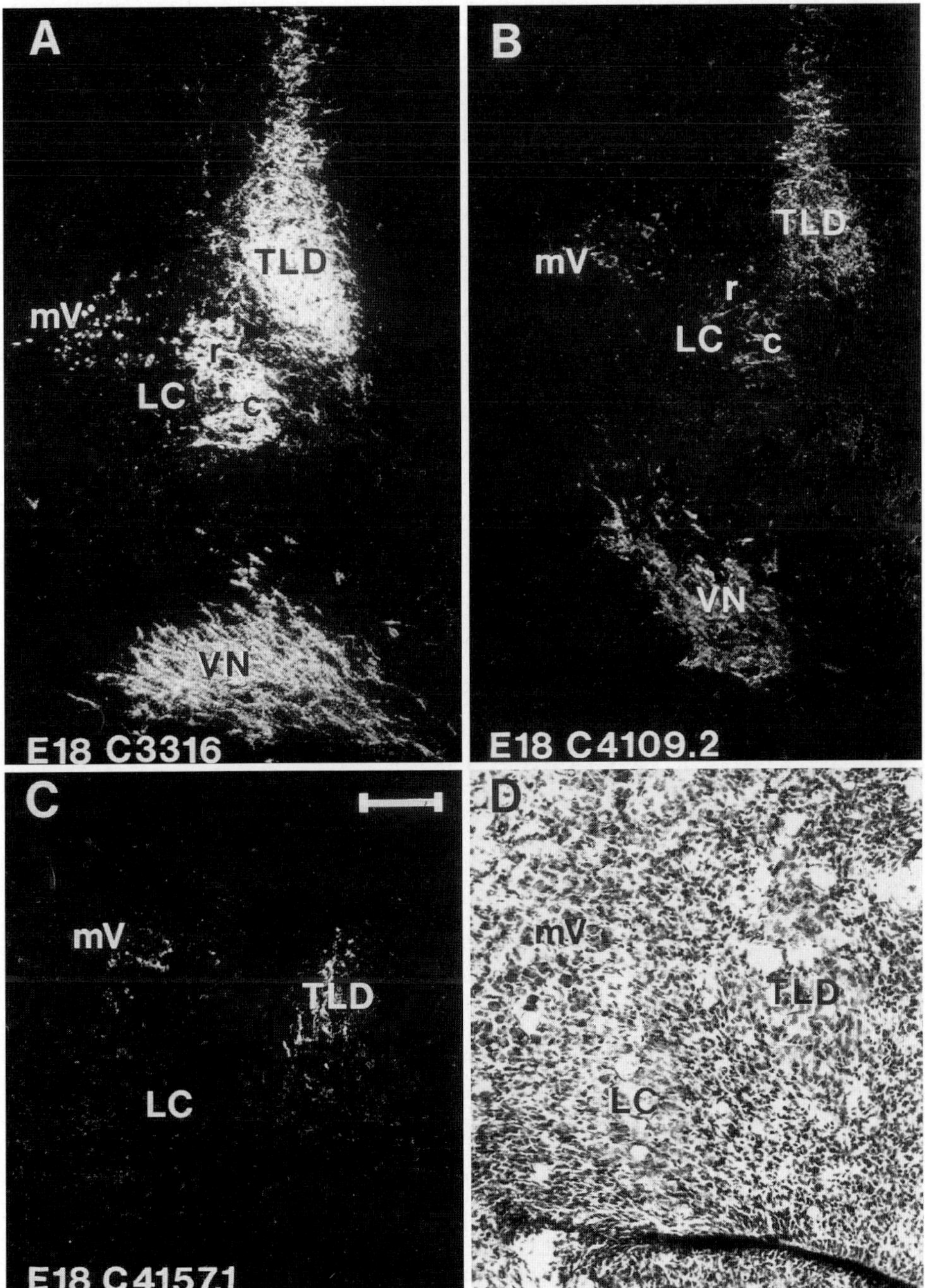

Fig. 28A–D. Retrograde labeling in nucleus tegmentalis laterodorsalis (*TLD*), locus coeruleus (*LC*) and mesencephalic trigeminal nucleus (*mV*) resulting from injections into cervical (**A**), mid-thoracic (**B**), and upper lumbar (**C**) spinal cord. TLD is labeled in all cases. The cervical injection results in labeling of rostral (*r*) and caudal (*c*) LC, while from the thoracic injection only cLC is labeled. **C** and **D** are adjacent sections; **D** is Nissl stained. LC and TLD are visible as *lighter stained nuclei* against a background of *small round cells*. *Bar*, 100 µm

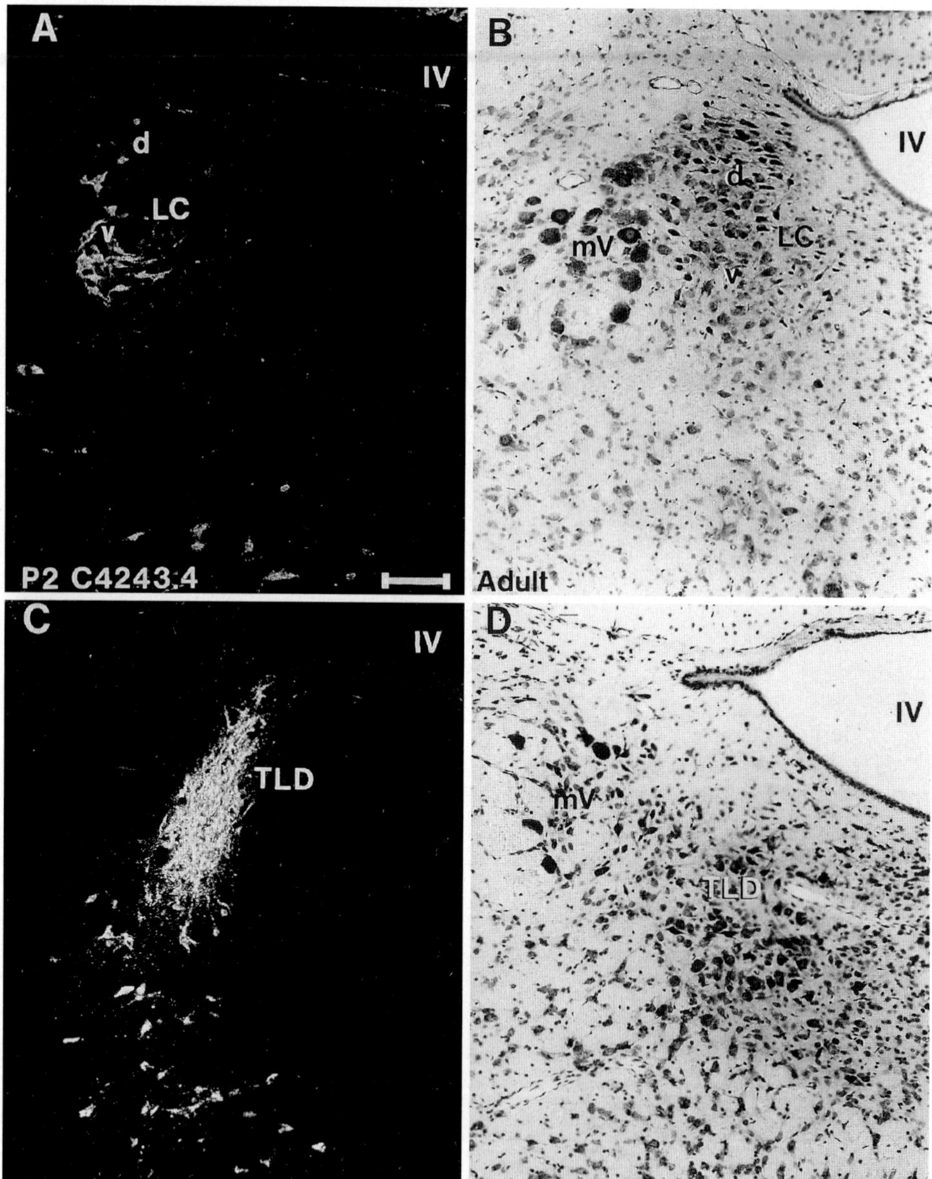

Fig. 29A–D. Adult-like labeling in both locus coeruleus (*LC*) and nucleus tegmentalis laterodorsalis (*TLD*) at postnatal day 2 (*p2*). Labeling is only present in the caudal and ventral (*v*) LC (**A**). Labeling of the TLD appears as a *tear-shaped* mass of fusiform cells (**C**). Section in **C** approximately 200 μm rostral from section in **A,B,D**, the LC and TLD are indicated in Nissl-stained sections of an adult rat brain. Note the relative positions of the TLD and LC with respect to each other (TLD medial to LC) and with respect to the mesencephalic trigeminal nucleus (*mV*; LC is immediately adjacent to the mV, while TLD is always separated from mV). *Bar*, 100 μm. *IV*, fourth ventricle

the gray matter of the lower lumbar spinal cord labeled only a few cells in the
ventral TLD (C4187.3). An injection into the same (lower) lumbar level at P2
(C4243.4; Fig. 29C,D) again resulted in retrograde labeling throughout the
TLD.

4.5.2 Nucleus Locus Coeruleus

From E18 onwards, the LC was easily distinguished (in neutral red-stained
sections) within the central pontine gray matter, due to the (optically) slightly
higher density of its constituent neurons. Before E18, the LC could only be
located relative to the nucleus mesencephalicus V and TLD: immediately
medial to the former and slightly caudal and lateral to the latter. The TLD
extended further rostrally than does the LC.

Due to the presence of the mesencephalic and pontine flexures, transverse
sections through the pons and rostral mesencephalon of younger fetuses resembled horizontal sections through these regions in the adult rat (Fig. 26).
Since this section plane made discrimination of the LC from the ventrally
located SC (in neutral red-stained sections) very hard, the transition from
central gray to pontine reticular nuclei immediately around the LC was used to
separate LC from SC. In the apparently horizontal section planes, the caudal
LC is located ventrally, and the rostral LC dorsally. The results of the retrograde labeling experiments are summarized in Table 3 (page 24).

At E17, an injection reaching up into the medulla oblongata (MO) retrogradely labeled neurons both in the rostral and caudal LC (C3523). From an
injection confined to the cervical and upper thoracic spinal cord, only neurons
located in the caudal LC were labeled (C4107.3), while an injection into the
lower cervical spinal cord labeled no LC neurons at all (C3531). In all E17
cases, only a few cells are labeled.

At E18, neurons in the rostral and caudal LC were labeled from injections
centered in the lower cervical spinal cord (C3316; Fig. 28A, C4162.3). Injections into the upper and middle thoracic spinal cord labeled caudal LC neurons
only (C4109.2; Fig. 28B, C4098.5). An injection into the lower thoracic and
upper lumbar spinal cord did not result in retrograde labeling of the LC
(C4157.1; Fig. 28C). Retrogradely labeled neurons in the caudal LC resulting
from a small injection confined mainly to the gray matter of the lower cervical
spinal cord indicated that the coeruleospinal fibers from the caudal LC entered
the cervical spinal cord white matter at E18 (C4162.3).

All E19 injections located in the cervical and thoracic spinal cord at E19
resulted in retrograde labeling of neurons in the caudal LC, but not in the
rostral LC (C3945.1, C3252, C3253, C4011.3, C4324.4, and C4324.6). No
retrograde labeling was obtained after a midlumbar injection (C3945.6).

From E20 onwards, all spinal cord injections located in both white and
gray matter resulted in retrograde labeling of neurons in the caudal LC, but
not in the rostral LC. At E21, an injection located mainly in the gray matter of
the lumbar spinal cord did not label neurons in the caudal LC (C4187.3). Since
this injection is also the caudalmost injection available (at this age), it is not
possible to discern whether fibers are just not present yet at this level of the

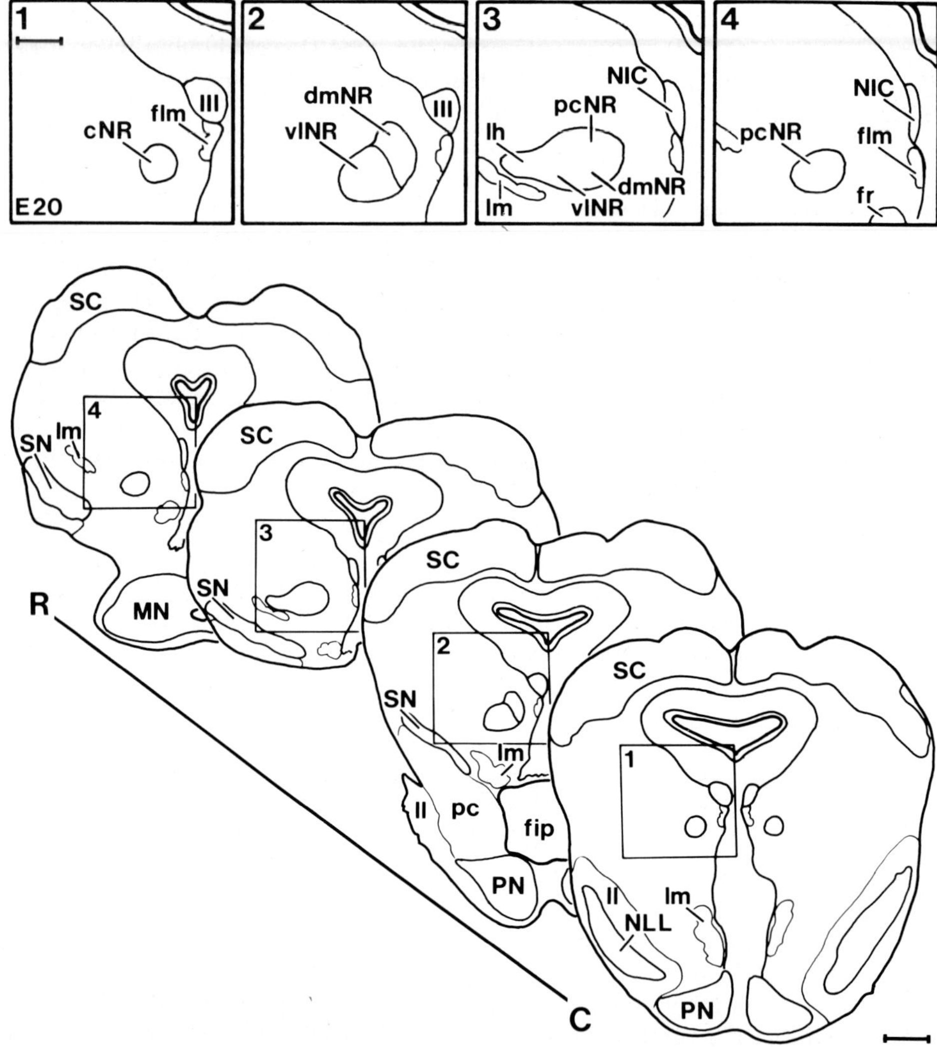

Fig. 30. Four diagrams were prepared (*top*) from four representative sections through the mesencephalon (*bottom*) for each age (embryonal day 20, *E20*, shown in this example). The diagrams are arranged from caudal (*C; 1*) to rostral (*R; 4*) Salient surrounding structures have been labeled in this figure only. Retrogradely labeled cells in the nucleus ruber (*NR*) were plotted into these diagrams (see Figs. 31, 32). *Bar* (*top left*), 100 μm; *bar* (*bottom right*), 200 μm. *cNR*, caudal pole of magnocellular NR; *flm*, longitudinal medial fascicle; *III*, oculomotor nucleus; *dmNR*, dorsomedial part of magnocellular NR; *vlNR*, ventrolateral part of magnocellular NR; *NIC*, interstitial nucleus of Cajal; *pcNR*, parvicellular part of NR; *lh*, lateral horn; *lm*, medial lemniscus; *fr*, retroflex fascicle; *SC*, nucleus subcoeruleus; *SN*, substantia nigra; *MN*, mamillary nuclei; *fip*, fossa interpeduncularis; *PN*, pontine nuclei; *ll*, lateral lemniscus; *NLL*, nucleus of lateral lemniscus

spinal cord or whether they have reached the injection level, but are confined to the white matter compartment of the cord. A similar injection into the same lumbar spinal cord level at P2, however, labeled only a few neurons in the LC (C4243.4; Fig. 29A,B).

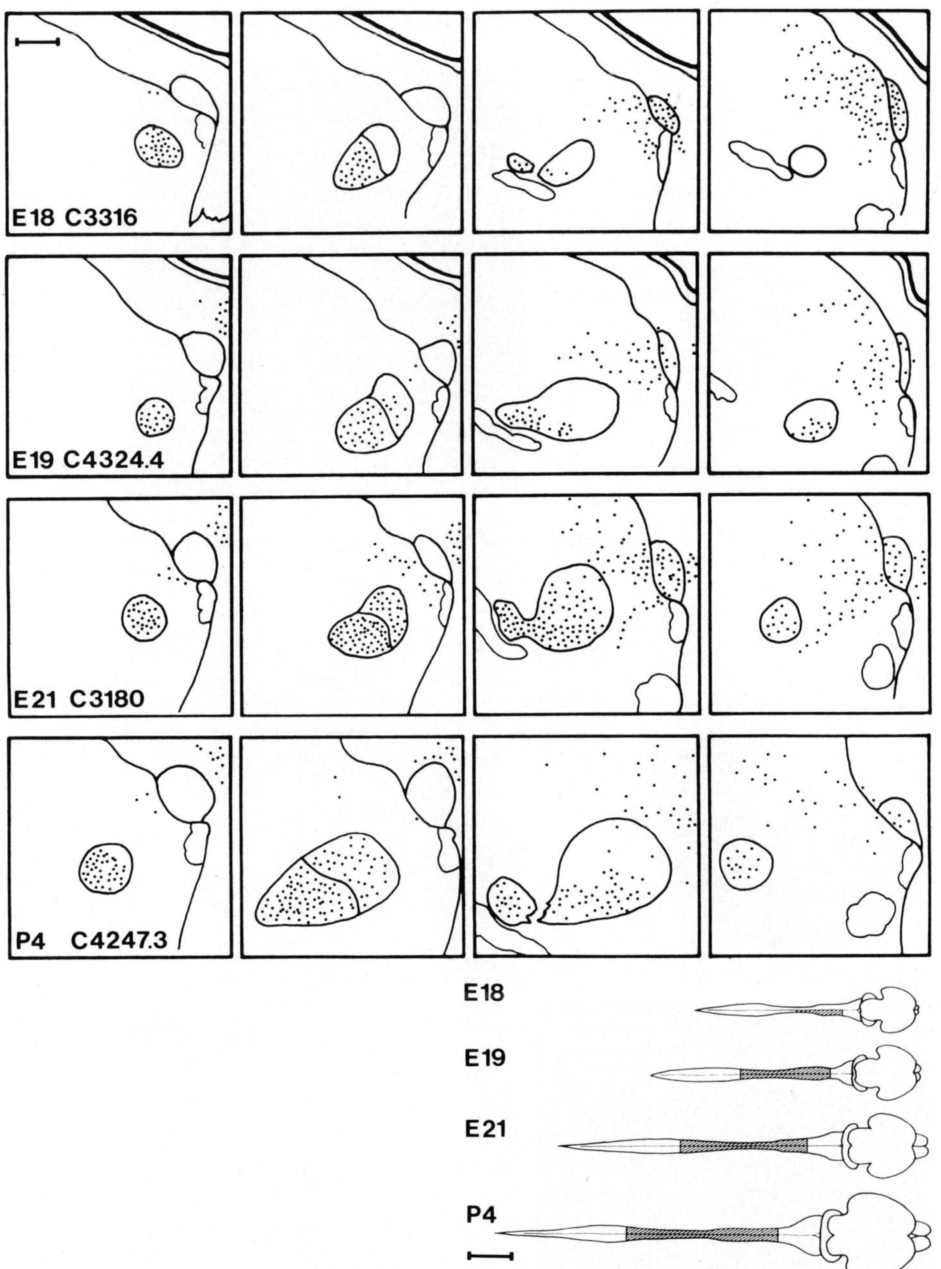

Fig. 31. Retrogradely labeled neurons from injections into the cervical intumescence at embryonal day 18 (*E18*), *E19, E21,* and postnatal day 4 (*P4*); *bar,* 100 µm. Diagrams of the pertinent sections are depicted (bottom; *bar,* 5 mm). The first fibers from the dorsomedial part of magnocellular nucleus ruber (dmNR) reach the cervical intumescence at E19, but it is not until E21 that retrogradely labeled cells appear throughout the dmNR. In the parvicellular part of nucleus ruber (pcNR), most labeled neurons are located rostrally at E19.

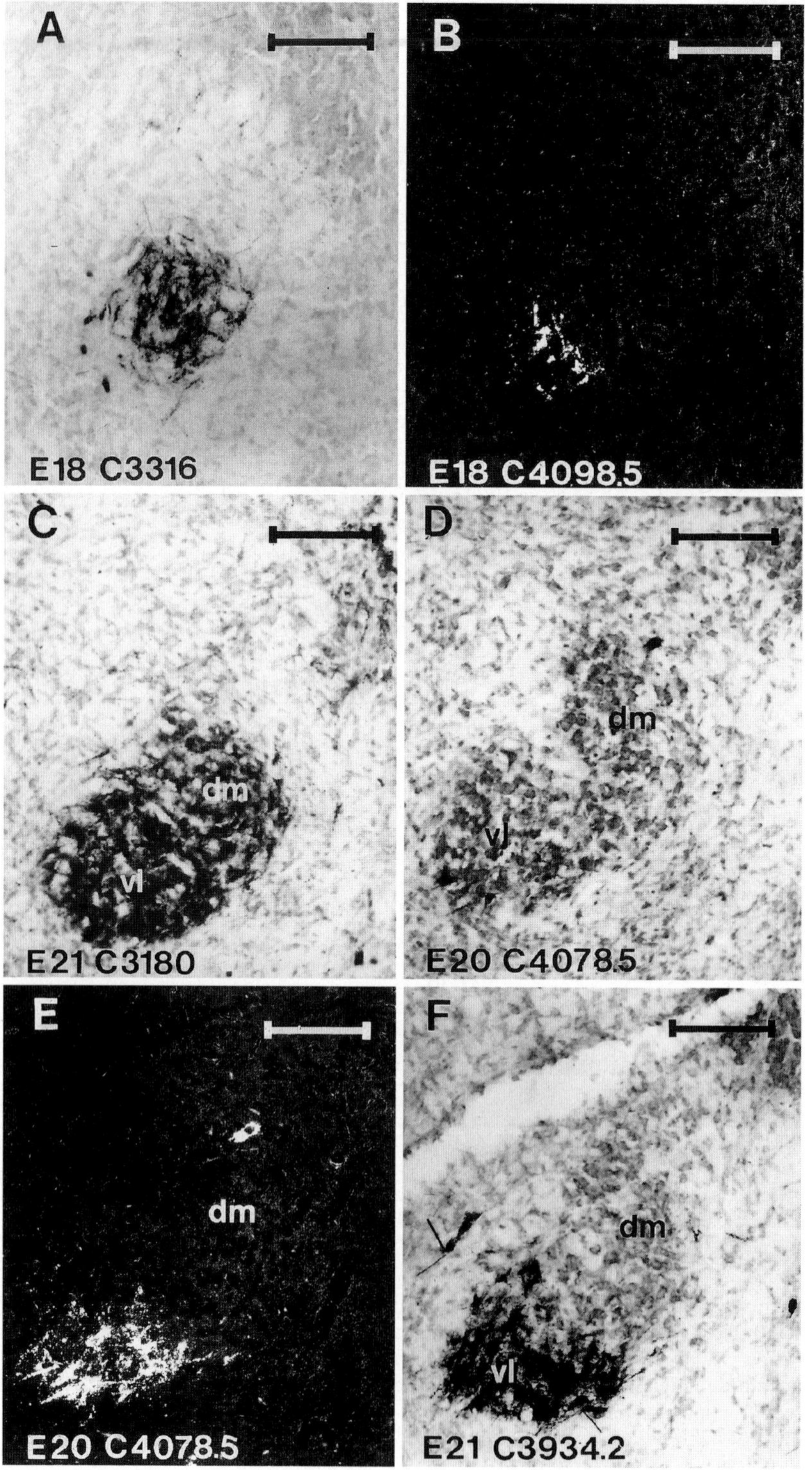
A
E18 C3316
B
E18 C4098.5
C
dm
vl
E21 C3180
D
dm
vl
E20 C4078.5
E
dm
E20 C4078.5
F
dm
vl
E21 C3934.2

4.5.3 Nucleus Ruber

The NR was easily distinguished at all ages examined as a relatively compact nucleus within the loosely packed neurons of the mesencephalic reticular nuclei. In order to describe these results, the NR was subdivided into cNR, vlNR, dmNR, and pcNR (Fig. 30; Reid et al. 1975). In most series, the NR was present in five to eight sections. The results of the retrograde labeling experiments are summarized in Table 3 (page 24).

At E17, the injection site in experiment C4258.6 reached up into the caudal pons. This injection resulted in closely packed, heavily labeled neurons in the cNR and vlNR. A few lightly labeled cells were observed in the caudal dmNR and the pcNR. In experiment C3523, the injection site extended well into the MO, but did not reach the pons. This resulted in lightly labeled neurons in the cNR and the vlNR. No labeled neurons were observed in the dmNR or pcNR. The injection site in experiment C4107.3 was confined to the cervical enlargement. In this case, the NR was devoid of labeled cells, though retrogradely labeled neurons were present in the nucleus interstitialis of Cajal and in the nucleus parafascicularis prerubralis, as in the two experiments mentioned above. In all other cases in which the injection sites were located further caudally in the spinal cord, retrograde labeling was absent from the NR.

At E18, an injection which had spread from the cervical enlargement into the thoracic spinal cord and into the caudal MO (C3315) resulted in heavy retrograde labeling of neurons in the cNR and vlNR. A few retrogradely labeled neurons were observed in the caudal dmNR. Some lightly labeled cells were also present scattered through the pcNR. In experiment C3316, the injection site was restricted to the cervical enlargement. This resulted in heavily labeled, closely packed neurons in the cNR and vlNR. Most labeled neurons in the cNR were located ventrolaterally (Figs. 31, 32A). In experiment C4109.2, the injection site was restricted to the thoracic spinal cord. In this case, only a few lightly labeled neurons were observed in the cNR and in the caudal vlNR. A large injection into the thoracic and upper lumbar spinal cord (C4098.5; Figs. 32B, 33) resulted in a few clearly labeled neurons in the ventrolateral cNR only, while a small injection into the same area (C4157.1) yielded only a few lightly labeled neurons in the cNR. Injections below the thoracic spinal cord (C4169.1 and C4162.1) left the NR free of retrograde

---

Fig. 32A–F. Retrogradely labeled neurons throughout the caudal (c) pole of magnocellular nucleus ruber (NR) from a cervical injection (**A**) and in the ventrolateral cNR only from a lumbar injection (**B**), both at embryonal day 18 (*E18*). The earliest descending fibers originate in the ventrolateral cNR. **C,F** At E21, a cervical injection resulted in retrogradely labeled neurons in both ventrolateral (*vl*)NR and dorsomedial (*dm*)NR (**C**), while a lumbar injection resulted in labeling of neurons in the vlNR (**F**). Fibers from the vl and dm subgroups have reached their projection levels in the lumbar and cervical spinal cord, respectively. A small injection into the lower cervical spinal cord resulted in retrogradely labeled neurons in the cNR and ventrolateral NR (**D,E**). This again demonstrates the orderly appearance of retrograde labeling in the NR along a caudal and ventrolateral to rostral and dorsomedial gradient. *Bars*, 100 μm

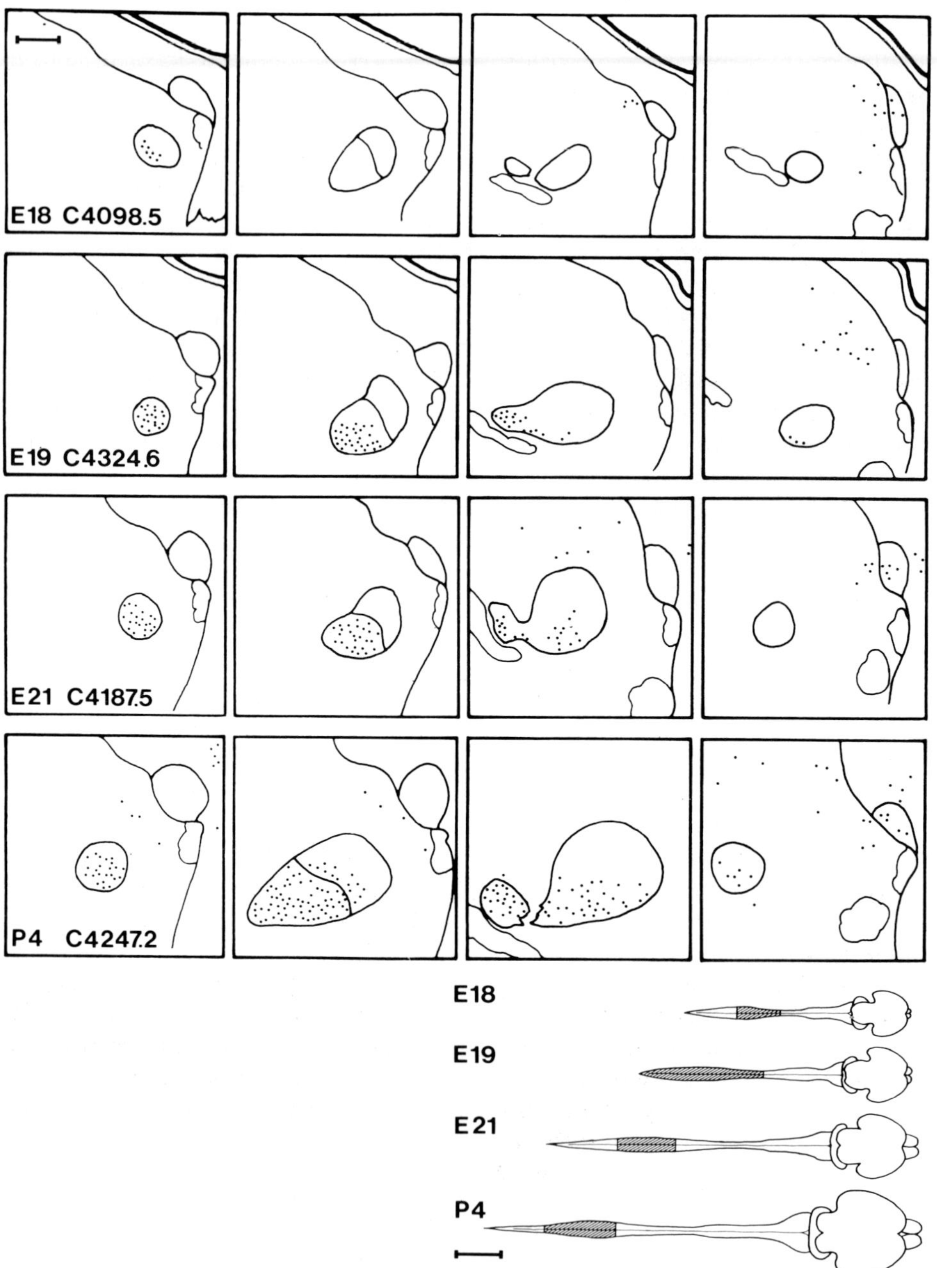

Fig. 33. Retrogradely labeled neurons from injections into the lumbar intumescence at embryonal day 18 (*E18*), *E19*, *E21*, and postnatal day 4 (*P4*); *bar*, 100 μm. Diagrams of the pertinent sections are depicted (bottom; *bar*, 5 mm). At E18, only neurons at the caudal pole of magnocellular nucleus ruber (cNR) were labeled from a lumbar injection, indicating that the earlier descending fibers originate from the cNR. Even at P4, labeling of the parvicellular part of nucleus ruber is sparse after lumbar injections

labeling In an experiment in which the injection site was restricted to the gray matter of the cervical spinal cord (C4162.3), no retrogradely labeled neurons were observed in the NR.

At E19, heavily labeled neurons resulted in the cNR and in the vlNR from a large injection which had spread up to the level of the obex (C3945.1) Many neurons, though only lightly labeled, were present in the dmNR and the pcNR. In general, the intensity of the labeling and the density of the labeled cells decreased ventrolaterally to dorsomedially. In experiments C3252, C3253, and C4324.4 (Fig. 31), the injection did not reach rostrally beyond the cervical enlargement. This resulted in retrogradely labeled neurons throughout the cNR (with a slight ventrolateral preponderance) and the vlNR. In the dmNR, labeled neurons were confined to its caudal half. Labeled neurons in the pcNR were located rostrally and ventromedially in experiments C3252 and C3253 and were absent in experiment C4324.4. After an injection into the lower thoracic and lumbar cord (C4324.6; Fig. 33), heavily labeled neurons were observed in the ventrolateral cNR and the vlNR. No labeled neurons were observed in the dmNR. An injection located in the midlumbar enlargement (C3945.6) did not label any neurons in the NR.

In experiment C4011.3, the injection was mainly restricted to the gray matter of the cervical enlargement. No labeled neurons were present in the NR, indicating that no rubrospinal fibers had invaded the gray matter yet.

In the E20 experiments C3657.2, C4255.2, and C4078.5 (Fig. 32D, E), the rostral border of the injection site was located approximately in the middle of the cervical enlargement. These injections resulted in retrograde labeling of neurons in the cNR and in the vlNR only. No labeled neurons were present in the dmNR. An injection into the middle thoracic spinal cord (C4273.5) resulted in light labeling of neurons in the cNR and vlNR, while a similar injection located in the lower thoracic and upper lumbar spinal cord (C4263.3) labeled only neurons in the cNR. An injection into the upper lumbar enlargement (C4263.4) did not label any neurons in the NR.

At E21, a large injection reaching up into the upper cervical enlargement (C3180; Figs. 31, 32C) resulted in heavily labeled neurons in cNR, vlNR, and dmNR. Both in the vlNR and the dmNR, almost all neurons appeared labeled, but the density of the labeled neurons was lower in the dmNR. Many labeled neurons were scattered throughout the pcNR. From injections into the upper lumbar enlargement (C3934.2; Fig. 32F, C4211.2, and C4187.5; Fig. 33), neurons were labeled in the ventrolateral cNR and the vlNR. No labeled neurons were observed in the dmNR or the pcNR. An injection located in the middle third of the lumbar enlargement (C4187.3) did not label any NR neurons at all.

Injections into the cervical enlargement at P2 (C4243.1) and P4 (C3208 and C4247.3; Fig. 31) resulted in retrograde labeling of neurons in all subgroups of the NR. A narrow crescent along the dorsomedial rim of the cNR remained free of labeled neurons. This non-labeled area was never observed in the prenatal experiments after injections located in the cervical enlargement.

Injections into the lower lumbar spinal cord at P2 (C4243.2, C4243.4), P3 (C4245.3, C4245.1), and P4 (C4247.1, C4247.2; Fig. 33) all resulted in retrogradely labeled neurons in the ventrolateral cNR and vlNR. Injections reaching up into the thoracic spinal cord at P3 clearly labeled neurons in the dmNR, apart from the expected retrograde labeling of neurons in the cNR and vlNR. Even in the P4 experiment C4247.2, which involved a lumbar injection, some retrogradely labeled neurons were present along the caudal and ventrolateral margin of the dmNR. At all postnatal ages and injection levels, retrogradely labeled neurons were present in the pcNR.

5 Discussion

In the first three parts of this chapter, we will discuss the time of appearance and the distribution patterns of the studied molecules in relation to the developmental processes that take place in the rat spinal cord (Sects. 5.1–5.3). In the fourth part, the development of several major fiber systems based on the intrauterine and postnatal neuronal tracing studies will be discussed (Sect. 5.4). In the following chapter, an attempt will be made to integrate the present results and to present new ideas on the development of the rat spinal cord. In order to focus properly on the events that take place in a particular layer of the spinal cord, the actual discussion of the results will be preceded by a short description of these developmental processes. Additionally, the influence and/ or regulatory action of the most relevant factors in these events will be described. Although our knowledge about these factors is incomplete and still subject to extensive study, the most important facts about the events of spinal cord development will be summarized. For reasons of clarity, the different regions that were identified within the layers will be dealt with collectively. Where necessary, a specific region or particular feature will be discussed separately if it appears to be the focus of local developmental events.

The sequence of the studied molecules will be consistent with that used in the previous chapter. First, the results of the studied structural compounds (neurofilaments and MAP) will be discussed. Second, results about the expression of SSEA-1 and AChE (functional compounds) will be presented. Third, the results of the studies about the distribution patterns of vimentin and GFAP (glial cell markers) will be dealt with. Finally, the results of the intrauterine and postnatal neuronal tracing studies will be discussed.

5.1 Matrix Layer

The matrix layer is the site of origin of the future spinal cord neurons and glial cells, i.e., the neuroblasts and glioblasts, respectively (Altman and Bayer 1984; Hirano and Goldman 1988). These cells are generated in the subventricular region of the matrix layer, which is situated adjacent to the membrana limitans interna surrounding the central lumen. In this relatively small region, repeated cell division leads to the formation of the pool of spinal cord neuroblasts and glioblasts. The actual proliferation of the subventricular cells is thought to start after the formation of the neural tube (Jacobson 1979). Nevertheless, it has been demonstrated that due to proliferation, the number of cells slightly increases in the still folding neural plate (Derrick 1937; Gillette 1944).

The neuroblasts in the spinal cord proliferate according to a ventral-to-dorsal and a rostral-to-caudal gradient (Sauer 1935; Fujita 1964; Nornes and Das 1974; Altman and Bayer 1984). So far, it is unknown whether the glioblasts of the spinal cord also proliferate according to a specific gradient.

After cell division, the future neurons extend cytoplasmatic processes towards the membrana limitans externa of the neural tube, and their nuclei migrate along these processes from the luminal surface towards the middle of the matrix layer. The mitotic cycle, including the migration of the nucleus along the cytoplasmatic extensions, is repeated several times until the processes are withdrawn from the membrana limitans externa. Just before the withdrawal, the nuclei return to the vicinity of the central lumen (Nornes and Das 1974; Altman and Bayer 1984). The exact role of this particular migration sequence is unclear, but it is thought that it allows regional cytoplasmatic factors to enter the nucleus to promote differential gene activity (see Jacobson 1979).

The spinal cord glial cells develop from two different ancestors, namely, the O2A progenitor cell, which differentiates into a type 1 astrocyte, oligodendrocyte, or a microglia cell (Raff et al. 1983), and the MARP cell, which differentiates into type 1 or 2 astrocytes (Fok-Seang and Miller 1991). Before differentiating into a more specialized glial cell, the O2A progenitors display a transient radial glial cell pattern (Hirano and Goldman 1988). Later in development, these radial glial cells developed into either a vimentin GFAP/GFAP-positive astrocyte or into a carbonic anhydrase-positive oligodendrocyte (Hirano and Goldman 1988).

[3H]Thymidine autoradiographic studies indicated that the different spinal cord neurons are generated in that part of the matrix layer corresponding to their future location in the mantle layer. For instance, the neuroblasts of the ventral matrix layer (or basal plate) are destined to become the motor neurons of the ventral mantle layer (or ventral horn). Similarly, the neuroblasts of the dorsal matrix layer (alar plate) will end up as the sensory neurons of the dorsal mantle layer (dorsal horn; Nornes and Das 1974). It has been proposed that according to the same principle, the ipsilaterally and contralaterally projecting neurons of the intermediate mantle layer (intermediate gray) originate in a separate region, intermediate between the ventral and dorsal parts of the matrix layer (Altman and Bayer 1984). So far, additional conclusive evidence for such a third intermediate generation zone is lacking.

The generation periods of the spinal cord neurons has already been established using [3H]thymidine autoradiography (Nornes and Das 1974; Altman and Bayer 1984). The future motor neurons are generated in a 2-day period, from E11 to E13. The larger motor neurons are generated ahead of the smaller ones (Kanemitsu 1971; Nornes and Das 1974; Altman and Bayer 1984). The future neurons of the intermediate gray are generated between E12 and E15, and the future interneurons between E14 and E16 (Nornes and Das 1974; Altman and Bayer 1984). In these periods, the cells also start to migrate towards their final location in the spinal cord mantle layer. So far, little is known about the generation period of the spinal cord glial cell types.

Two special anatomical features in the rat spinal cord matrix layer are the ventrally situated floor plate and the dorsally located roof plate. During recent years, our knowledge about these structures has expanded. Earlier, they were thought to be the source of spinal cord glial cells, and the resiliency of these cells was thought to result in the formation of the dorsal and ventral midline structures (see Altman and Bayer 1984). Recently, it was demonstrated that the floor plate is an important source of inductive signals (see Glover 1991). Apart from its specific chemotropic activity, which is thought to attract or repel axons to the midline (Tessier-Lavigne et al. 1988), the floor plate is also believed to induce developmental processes within the motor neurons in the ventral mantle layer. Of interest with regard to its development is the fact that a second floor plate-like structure can be induced by the implantation of notochord tissue (van Straaten et al. 1985, 1988). Additionally, a direct or indirect influence by the notochord on the proliferation of the basal plate neuroblasts has been suggested (van Straaten and Drukker 1987; van Straaten et al. 1989).

The roof plate of the matrix layer gradually develops from a wedge-shaped structure into a thin scptum. So far, it is thought to function as an axon barrier which regulates the course of the commissural and dorsal column axons. Several compounds related to axon guidance were found in the roof plate, but their direct relation to a barrier function is still uncertain (Snow et al. 1990).

Of the three different layers in the rat spinal cord, the developmental events that occur in the matrix layer are probably the best known. In contrast to the two other layers, the developmental processes of the matrix cells are more easily to manipulate in vitro. Consequently, the influence or regulatory role of different factors on several of these developmental events, i.e., induction, proliferation, and migration, has been extensively studied.

Both the neuroblasts and the glioblasts are generated from multipotent germinal cells. So far, the mechanisms that regulate the proliferation of the apparently homogeneous population of germinal cells (also designated neuro-epithelial cells, stem cells, or pre-progenitors) are largely unknown. One way of producing different kinds of cells from one type of ancestor is by a progressive restriction in its genetic capacity. Such a mechanism has been demonstrated before in the retina (Hicks et al. 1959, 1961; Hicks and D'Amato 1963, 1966). Recently, it was demonstrated that transiently expressed gene transcripts are capable of regulating the pattern formation in mesodermal tissue. Additionally, it was concluded that these genes were activated in response to an epigenetic, mesoderm-inducing growth factor (Christian et al. 1991). This particular study supports the idea that the asymmetric development of the spinal cord stem cells is under intrinsic (genetic) as well as strong extrinsic (epigenetical) influence.

Several studies, most of them in vitro, have shown the existence of a vast group of molecules that either regulate or influence the proliferation of the different central nervous system cell types. Neuronal proliferation, for instance, is stimulated by basic fibroblast growth factor (Gensburger et al. 1987). This particular growth factor is produced by the neuroblasts themselves, which implies an autocrine mechanism (Gensburger et al. 1987). The proliferation of astrocytes is stimulated by acidic as well as basic fibroblast growth factor (Pettmann et al. 1985). Both astrocyte and oligodendrocyte proliferation is regulated by a 20-kDa human B cell growth factor (Ecclestone and Silberberg

1985; Benviniste et al. 1989). Other factors that are believed to play a developmental role in the central nervous system are nerve growth factor, brain-derived neurotrophic factor, and, especially with regard to the proliferation of matrix cells, neurotrophin-3 (Maisonpierre et al. 1990).

The above-mentioned studies all demonstrated that as well as a genetic program, the cells of the matrix layer need extrinsic signals provided by a group of epigenetic factors that are still growing. An elegant study carried out several years ago demonstrated this genetic–epigenetic influence on cell proliferation by showing that cultured matrix cells require living cells for the environmental signal(s) influencing their proliferation (Temple 1989). The same study also showed that despite a standardized culture environment, different clones were still produced by the matrix cells, demonstrating their variable intrinsic program.

The production of the spinal cord glial cells also appears to be under strict epigenetic influence. In addition to the influence of factors that are able to act from greater distances, other more direct means of communication were also demonstrated. In vitro experiments revealed that during gliogenesis, cell–cell contact frequently occurs, suggesting direct transfer of soluble factors that influence proliferation (Temple and Raff 1985). It was found that type 1 astrocytes stimulate the O2A (glial) progenitor cells by means of a soluble factor (Noble and Murray 1984), which leads to an increased number of immature oligodendrocytes (Bhat and Pfeiffer 1986). Part of the O2A progenitor cells, however, remains in an indifferent state and continues to divide to replenish the progenitor number (Hardy and Reynolds 1991). Gliogenesis was also found to be influenced by neurons, which promote not only the differentiation of the oligodendrocytes (Fulcrand and Privat 1977; Valet et al. 1978, 1983), but also their survival (David et al. 1989).

Between E12 and E17, the present study demonstrated the temporary presence of MAP 1A and MAP 5 in a typical radial staining pattern in the matrix layer. Both proteins appeared and disappeared according to a ventral-to-dorsal gradient. So far, these particular MAPs were not thought to be involved in developmental events in the spinal cord matrix layer. Previously, it was established that between E11 and E16, the three types of spinal cord neurons proliferate in the matrix layer according to a ventral-to-dorsal gradient (Nornes and Das 1974; Altman and Bayer 1984). During this period, the nuclei of the neuroblasts migrate several times towards the midst of the matrix layer and back to the central lumen along cytoplasmatic processes extended earlier. The entry of regional cytoplasmatic factors into the migrating nucleus is supposed to regulate specific gene activities (Jacobson 1979). As a possible guide for the nuclei migrating back and forth, it would be most suitable for the cytoplasmatic extensions to be equiped with a dynamic cytoskeleton. In fact, in such a role a rigid skeleton seems unworkable. Both MAP 1A and MAP 5 are known for their role in the assembly and stabilization of microtubules (Riederer et al. 1986; Kuznetsov et al. 1981), which could serve the dynamics of the cytoskeleton. Thus, it seems acceptable that the transient, radial staining pattern in the matrix layer reflects the involvement of MAP 1A and MAP 5 in the organization of a dynamic cytoskeleton in the cytoplasmatic extensions of the proliferating neurons.

The exact function of the cytoskeleton in the neuroblast extensions is unclear. Most obvious is the above-mentioned role as a (dynamic) backbone of the cytoplasmatic extensions along which the neuroblast nucleus migrates during the mitotic cycles. However, the peculiar behavior of the neuroblasts during their mitotic cycles in particular, suggests the idea of a more elaborate role. It may be that the cytoskeleton participates in the nutrition and maintenance of the contacts of extensions with the internal and external membranes of the neural tube. With regard to this possible function, it was shown earlier that tubulin, the basic unit of microtubules and an important requirement for neuronal maturation (Yamada et al. 1970), is also present in the spinal cord matrix layer in a radial distribution pattern comparable to that demonstrated for MAP 1A and MAP 5 in the present study (Tucker et al. 1988). Also of interest is the finding that neuraxin, a molecule closely related to MAP 5, may be implicated in neuronal membrane–microtubule interaction (Rienitz et al. 1989).

Although it may seem surprising that structural morphoregulators, as described and discussed above, are actually present in the developing rat spinal cord matrix layer, it may be more intelligible if one considers the developmental events that take place in this layer. The particular processes, i.e., cell proliferation and migration, may need the active or passive participation of the cytoskeleton. Therefore, those structures that constitute and/or regulate the composition of the cytoskeleton are expected to be present in the developing matrix layer. Their obscurity may be explained by the fact that so far only a few studies have focused on the presence of structural compounds in the spinal cord matrix layer, and even fewer on their function.

Between E13 and E18, MAP 1A is present in the floor plate. A patch-like presence of MAP 5 was detected in the floor plate between E12 and P4. It could not be determined whether the proteins were located inside or outside the floor plate cells. Their temporary presence suggest a developmental role for both proteins, but so far it is unclear what kind of function they could play. The floor plate is thought to play a role in attracting or repelling developing (commissural) axons (Dodd and Jessel 1988; Tessier-Lavigne et al. 1988). Although both MAP 1A and MAP 5 are present in the floor plate during the period in which commissural axons "search" their pathway, it seems unlikely that both proteins play a role in this guiding process. Future studies should reveal the function of MAP 1A and MAP 5 in the floor plate of the rat spinal cord.

Rat embryos between E9 and E15 express SSEA-1 in the subventricular region of the matrix layer. In addition, between E13 and E17, this particular antigen was also present in the dorsal part of the matrix layer. These findings are in agreement with earlier reports (Marani and Tetteroo 1983; Marani 1986; Yamamoto et al. 1985; Oudega et al. 1992). In the subventricular region, the neuroblasts divide before migrating towards the midst of the matrix layer. The present results, therefore, indicate a twofold involvement of SSEA-1 in the development of the neuroblasts: firstly, a general involvement in neuroblast division, and secondly, a specific role in the migration of the dorsal neuroblasts towards the periphery of the matrix layer. The involvement of SSEA-1 in the multiplication of neuroblasts is underlined by the fact that later in the proliferation phase, when cell division has ceased, the subventricular layer appeared to be devoid of SSEA-1. The exclusiveness of SSEA-1 for those neuroblasts that are destined to settle in the dorsal horn of the spinal cord demonstrates a special relationship between the antigen and the future interneurons of laminae II and III (see also Sect. 5.2). Such a relationship has been suggested before in the rat (Dodd and Jessel 1985; Jessel and Dodd 1985) as well as in the quail (Sieber-Blum 1989).

The precise function of SSEA-1 is still obscure, but it has frequently been implicated in cell–cell contact-mediating processes. Through cell–cell contacts, stage-specific carbohydrates were found to be capable of regulating biological processes (Solter and Knowles 1978; Rastan et al. 1985; Niedieck and Löhler 1987; Eggens et al. 1989). Possible functions of SSEA-1, a stage-specific carbohydrate, and implications of the expression patterns in the developing matrix layer descibed above will be discussed more extensively in chap. 6.

The expression of SSEA-1 on glial cells has been described previously (Niedieck and Löhler 1987), something which suggests that SSEA-1 distribution in the dorsal matrix layer reflects the presence of glial cells. However, with respect to the distribution pattern as described in the present study, it seems unlikely that SSEA-1 is expressed on glial cells in the dorsal matrix layer. The dorsally located glia appears as radially orientated, vimentin-positive cells (see below). This pattern does not correspond to the overall distribution of SSEA-1 as found in the developing alar plate.

During the early days of matrix layer development, AChE was found to be present in its ventral part, which suggests an exclusive presence of the enzyme in proliferating (motor) neuroblasts (see also Oudega and Marani 1990). The enzyme has rarely been associated with proliferating neurons of the central nervous system. Recently, a small group of "primitive" choline acetyltransferase (ChAT)-positive cells was demonstrated around the ventral half of the mantle layer (Phelps et al. 1991). More importantly, however, photographs of the spinal cord of E12 rat embryos in the same report clearly revealed the presence of ChAT in a population of cells within the ventral matrix layer. This result, which was not discussed in the paper, supports the present finding of a group of cells with a cholinergic character in the ventral matrix layer of the rat spinal cord and indicates that the future cholinergic neurons of the rat spinal cord are already committed to their phenotype in an early (premigratory) stage of their development. A possible function of the enzyme in the future spinal cord motor cells could be an involvement in the control of the final phase in their proliferation. Due to the exclusiveness of the present results, a more in-depth discussion on the implications of the AChE presence in the matrix layer is given in Chap. 6.

Around E16, the enzyme-positive ventral region (basal plate) was sharply delineated from the enzyme-negative dorsal part (alar plate) of the spinal cord matrix layer. The border between the basal and alar plate was found at the level of the sulcus limitans. This could indicate that the development of all the ventral cells of the spinal cord matrix layer is at least partly regulated by AChE. Earlier, the existence of a third generation zone within the spinal cord matrix layer, intermediate between the basal and alar plate and producing the future relay neurons, was proposed (Altman and Bayer 1984). So far, however, conclusive evidence regarding such an additional generation zone is lacking. Also, the present histochemical results do not support the proposal by Altman and Bayer (1984); they demonstrate a twofold subdivision of the matrix layer in an AChE-positive basal plate and a SSEA-1-positive alar plate (see also Chap. 6). However, it is possible that the future spinal cord relay neurons also require AChE for their proper development and that, consequently, this third generation zone does not stand out as a separate area after AChE staining.

The results mentioned and discussed above about the two studied functional morphoregulators SSEA-1 and AChE clearly demonstrate that the spinal cord matrix layer contains histochemically different cell populations. Apparently, the matrix layer should not be considered as a homogeneous population of cells, as has been thought for a long time. The presence of histochemically distinguishable cell populations seems acceptable in view of the different developmental pathways of the different types of matrix layer cells (see also Chap. 6).

The present results demonstrated a scattered pattern of the glial cell marker vimentin in the matrix layer of the rat spinal cord. Later on, a more regular radial pattern was found to develop. Such a regular vimentin pattern appeared to be present in the rat spinal cord matrix layer roughly between E11 and P4. After P10, vimentin was found to be expressed in short, twisting fibers that emerged from the matrix layer into the mantly layer. The replacement of vimentin by the other studied glial cell marker GFAP took place during the first 3 postnatal weeks. After this transition period, GFAP-positive fibers, originating in the intermediate aspect of the matrix layer, curved sharply and coursed into either the ventral or the dorsal midline structure.

It has been demonstrated that during gliogenesis, the O2A progenitors develop into either vimentin-positive type 1 astrocytes or carbonic anhydrase-positive oligodendrocytes (Hirano and Goldman 1988). In vitro and in vivo, the O2A progenitors were found to express vimentin even before their differentiation into type 1 astrocytes (Hardy and Reynolds 1991). Therefore, the early scattered vimentin positivity demonstrated in the spinal cord matrix layer most likely reflects the presence of the glia cell progenitor.

Yet another explanation of the (early) scattered vimentin pattern could be the presence of vimentin in neuroblasts rather than in the glial cell progenitors. Earlier, an in vitro coexpression of vimentin with neurofilament subunits in developing neurons was demonstrated, which could indicate that vimentin plays a role in neurogenesis (Bignami et al. 1982; Cochard and Paulin 1984; Lee and Page 1984). It is also thought that cytoplasmatic factors are involved in the regulation of specific gene activities (Jacobson 1979). Also of interest with respect to the above-mentioned result is the proposed temporal, more general, relationship between vimentin and neural tube development (Houle and Federoff 1983).

Later on in development, the regular, striped appearance of vimentin in the rat spinal cord matrix layer could reflect the presence of glial extensions that support the final radially orientated migration of the spinal cord neuroblasts towards the mantle layer (see Leber and Sanes 1990). An orthogonal migration pattern of the spinal cord neuroblasts within the matrix layer was suggested previously (Altman and Bayer 1984). Although often referred to and described, it was unclear for a long time whether migrating neuroblasts were actually associated with the radially orientated glial processes. Recently, it was demonstrated that the neuroblasts follow the radially orientated extensions in order to migrate through the matrix layer to reach their final destination in the mantle layer (Gasser and Hatte 1990). They place their cell body against the glial fiber and extend a motile, leading process in the direction of the migration. This (neuronal) extension was found to enfold the glial fiber with short filipodia and lamellipodia (Gasser and Hatte 1990). The time period in which vimentin could be detected in the matrix layer (E11–P4) overlaps with the period in which the neuroblasts are known to migrate towards the periphery of the matrix layer to form the mantle layer (E11–E17). An even more convincing reason for such a structural support provided by the vimentin-positive radial glial fibers to the migrating neuroblasts is the fact that the vimentin fibers appear according to a ventral-to-dorsal gradient, which in terms of direction and time is similar to the gradient according to which the neuroblasts are generated. Specifically, the presence of vimentin in the cytoskeleton of the spinal cord glial cells could contribute to their tensile strength needed for the structural support of the neuroblasts. On the other hand, other roles have been proposed for vimentin, e.g., in DNA replication/recombination and in gene expression (Geiger 1987; Georgates and Blodel 1987a,b), all of which could indicate a more basic regulatory function for vimentin in the development of the spinal cord glial cells.

Especially in the floor plate, abundant vimentin presence could be detected during the early development of the matrix layer (E13–E15). After E15, the regularly distributed, vimentin-positive, fiber-like structures gradually disappeared from the floor plate. A direct link with the development of the ventral raphe could not be made, as the fibers of the ventral midline originated in the intermediate matrix layer (see below). Future studies are necessary to elucidate the role of the temporary abundant presence of vimentin in the floor plate.

The presence of GFAP in the rat spinal cord matrix layer is restricted to the ventral and dorsal midline structures. Therefore, its presence and possible role as well as that of vimentin in these raphe-like structures will be discussed in relation to the mantle layer (see below).

In the present study on the expression of glial cell markers in the developing rat spinal cord, two intriguing findings were noticed. At E11–E12, vimentin could already be found in the membrana limitans externa. Short, vimentin-positive protrusions penetrated from this membrane into the most ventral part of the matrix layer. The precise localization of these vimentin-positive protrusions should be established with the help of electron microscopy, which could then lead to the elucidation of their function. The second peculiar finding was the presence of tangles of vimentin-positive fibers found in the matrix layer during the early development. Their origin and function is intriguing and remains to be elucidated

The above-mentioned results and discussion on the presence of the two glial markers vimentin and GFAP in the rat spinal cord matrix layer indicates an elaborate role for vimentin and a more restricted role for GFAP in the different developmental events. So far, the presence and functions of vimentin and GFAP in the developing rat spinal cord have only be sparsely studied.

Additional information about these molecules is lacking and, consequently, the presented possible roles are to be considered as highly speculative and merely as suggestions for future investigations.

5.2 Mantle Layer

At the end of the mitotic phase, which is marked by a decrease in DNA synthesis, the neuroblasts migrate towards the periphery of the matrix layer to form the mantle layer. The migration of the neuroblasts is thought to be initiated by their withdrawal from the mitotic cycles, although some reports mentioned a continuing proliferation during their migration (Sauer 1959; Cowan 1978). In general, the direction of migration is orthogonal to the ventricular surface. As described above, it was unclear whether close association between migrating neuroblasts and radially orientated glial processes existed. Only recently it was actually demonstrated that the neuroblasts place their cell body against a glial fiber and then extend a motile, leading process in the direction of their migration. This (neuronal) extension was found to enfold the glial fiber with short filipodia and lamellipodia (Gasser and Hatte 1990).

Even before the start of migration, the final location of the cells in the mantle layer seems to be determined. The cells settle down in anlagen, which consist of those neurons that together constitute a functional, integrative circuit. Cells that assemble in an anlage are largely similar, although differences among them do exist. It seems that morphologically and chemically identical cells or those with a common functional purpose are capable of recognizing each other and of gathering in anlagen. The settling neurons then extend their axon and dendrites and finally establish the proper connections (see Sect 5.3; Van der Loos 1965; Purves and Hume 1981; Bray 1982).

Earlier, it was demonstrated that the different types of neurons in the spinal cord mantle layer originate in the corresponding region of the matrix layer (Altman and Bayer 1984). Accordingly, the ventrally situated motor neurons are generated in the ventral matrix layer (basal plate). In the same way, the dorsally situated interneurons originate in the dorsal part of the matrix layer (alar plate; Nornes and Das 1974). It has also been proposed that the ipsilaterally and contralaterally projecting neurons of the intermediate region of the mantle layer (intermediate gray) are generated in a separate area of the matrix layer, intermediate between the basal and alar plate (Altman and Bayer 1984). So far, conclusive evidence for such a third generation zone is lacking. More recently, it was demonstrated that the developing matrix layer contained at least seven separate groups of cells located along the ventricular surface. This finding suggests that the neurons of the mantle layer originate in several different zones in the matrix layer (Silos-Santiago and Snider 1990).

Between E12 and E17, the motor neurons were found to settle down in the motor neuron columns in the ventral horn (Angulo Y González 1990). Traditionally, these columns are referred to as Rexed's lamina IX (Rexed 1954). It was found that the larger motor neurons are generated ahead of the smaller ones (Nornes and Das 1974; Altman and Bayer 1984). Between E12 and E18, the motor cells extend their axon and dendrites (Altman and Bayer 1984).

The ipsilaterally and contralaterally projecting (relay) neurons arrange themselves in the intermediate gray (Rexed's laminae IV and VIII; Rexed 1954) and in the most dorsal layer of the dorsal horn (Rexed's lamina I; Rexed 1954). Around E12–E13, the first contralaterally projecting axons are present in the ventral commissure and ventral funiculus of the rat spinal core (Altman and Bayer 1984; Oudega et al. 1990a). From E13 on, a special group of cells known as the intermediolateral cell column was found to migrate towards the lateral part of the intermediate gray (lateral horn). In addition, from E16 on the cells that form the intermediomedial cell column were found to migrate towards the central canal (see Altman and Bayer 1984). These two cell groups will be discussed separately.

From E14 on, the dorsal horn is occupied by the spinal interneurons (Nornes and Das 1974; Altman and Bayer 1984). These neurons are arranged in Rexed's laminae II and III (Rexed 1954). Most of the dorsally located interneurons receive nociceptive information from the periphery of the body. A large group of compounds were found to be involved in the processing of the peripheral nociceptive information (for review see Blumenkopf 1988), among them substance P (Kuraishi et al. 1985).

In general, the development of the rat spinal cord mantle layer, i.e., the migration of the neurons and their settling in the anlagen, follows a ventral-to-dorsal gradient (see Altman and Bayer 1984). Most likely, the migration of the cells is not only regulated by the radially orientated glial fibers (see Gasser and Hatte 1990), but also strongly influenced by chemical gradients in combination with, or solely by, differential adhesion to the environment (Holtfreter 1944; Moscana 1952). The process of migration starts in the matrix layer and continues in the mantle layer. Therefore, several factors that influence the cells in the matrix layer still seem to be involved with these cells after they have entered the mantle layer. In addition, other factors are known to be involved in more specific developmental events that take place in the spinal cord mantle layer. For instance, the initial outgrowth of the neurites of the mantle layer neurons is promoted by fibroblast growth factor (Morrison et al. 1986; Unsicker et al. 1986; Walicke et al. 1986). The same factor appears to stimulate the maturation of the astroblasts into astrocytes (Pettmann et al. 1985) and the maturation of the spinal cords oligodendrocytes (Ecclestone and Silberberg 1985). Another factor apparently involved during development is the so-called brain-derived neuronal factor (Maissonpierre et al. 1990).

Besides the factors that seem to be involved in the continuation of the cellular differentiation processes, other molecules are known to be involved, in particular, in the migration of the neurons towards their proper location. Important in this respect is the group of extracellular substrate adhesion molecules, including the well-studied neuronal cell adhesion molecule (N-CAM). In general, these molecules regulate cell–cell contacts and, in doing so, they ensure synchronized interaction (Edelman et al. 1990). Another category of regulating compounds is best represented by cytotactin, a repulsive agent that is highly involved in the determination of the migratory pathways of neurons (Davies et al. 1990; Raper and Kafhammer 1990). Finally, the maintenance of the mantle layer neurons also seems to be influenced by external factors. For instance, the in vitro survival of the spinal cord motor neurons is increased by factors present in a conditioned medium of their targets, the peripheral muscles (Martinou et al. 1989).

In general, the distribution patterns of the studied cytoskeleton proteins in the spinal cord mantle layer appeared to be largely similar. The molecules were more prominently present in the ventral horn than the intermediate and dorsal regions of the mantle layer. Overall, within the developing mantle layer, neurofilaments were faintly expressed, whereas the MAP were found to be abundantly present. The faint presence of neurofilaments most likely reflects the specificity of the antibody used (NF-90), which recognizes the phosphorylated forms of the neurofilament subunits (Oudega et al. 1990a). Phosphorylation of the subunits is known to take place after they have entered the axons (Carden et al. 1985; Georges et al. 1986; Oblinger 1987). From a functional point of view, subunit phosphorylation is considered to be a means by which the neurofilaments cross-link and stabilize the axonal cytoskeleton (Hirokawa et al. 1984; Matus 1988; Nixon and Sihag 1992). Nevertheless, the present study demonstrated that during early embryonal development (E12–E16), phosphorylated neurofilaments are also expressed in the ventrally situated motor neurons and their dendrites. The presence of phosphorylated neurofilaments in the spinal cord motor neurons has been demonstrated before with antibodies that recognize partially or completely phosphorylated neuro-filaments (Carden et al. 1985). So far, the function of these phosphorylated subunits in the motor cells is unclear. It could be that they represent NF-L subunits being phosphorylated to prevent early assembly and/or to disassemble already formed filaments (see Nixon and Sihag 1992).

Around E12–E13, both MAP 1A and MAP 5 are localized in cells and fibers of the ventral horn. During the same period, both antigens were also present in fiber-like processes that course in the mantle layer close to the matrix layer and in the ventral commissure. These distribution patterns in-dicates an involvement of MAP 1A and MAP 5 in the regulation of the microtubular organization in the developing motor neurons and their axons and dendrites as well as in the contralaterally projecting axons of the rat spinal cord. Around E12, the motor neurons of the ventral horn are known to start the outgrowth of their axon and dendrites (Altman and Bayer 1984). Around the same time, the first contralaterally projecting axons were found to be present in the ventral commissure and ventral funiculus (Altman and Bayer 1984; Oudega et al. 1992a). Both MAP 1A and MAP 5 are known for their role in the assembly and stabilization of microtubules; hence, they are especially active in growing neuronal extensions (Kuznetsov et al. 1981; Riederer et al. 1986; Ferreira et al. 1989). Although the early expression of MAP 1A is unexpected (see Oudega et al. 1990b), MAP 5 is generally accepted as an early MAP and thought to be involved in neurite outgrowth processes (Riederer et al. 1986). The present results demonstrate an involvement in the development of the early rat spinal cord morphology for both MAP. The findings underline the developmental role for MAP 5 and additionally suggest such a function for the MAP 1A molecule. Consequently, in relation to rat spinal cord develop-ment, both MAP 5 and MAP 1A are to be considered as early MAP.

With regard to their presumed function in the assembly and stabilization of microtubules, the distinct presence of MAP 1A and MAP 5 in a subpopulation of adult motor neurons is surprising. One should not expect these two "developmental" events to be ongoing in adult neurons, at least not at rates that require large amounts of the MAP. A possible explanation for the immunoreactivity could be that it reflects the basic turnover of both molecules. However, if this is the case, one would not expect the proteins to be abundantly expressed in the neuronal cell bodies. Clearly, the exact role of MAP 1A and MAP 5 in the adult motor neurons remains to be elucidated. One possibility is that both MAPs are involved in the plasticity of the neuronal processes of the motor neurons (see next paragraph).

The overall faint appearance of MAP 1A and MAP 5 immunoreactivity in the postnatal and adult spinal cord mantle layer indicates the presence of low amounts of the proteins. The presence of these MAPs, which are known to be involved in microtubule assembly, may benefit the plasticity of adult spinal cord axons. If one assumes that the adult spinal cord neurons do have the capacity to react to certain changes in their morphology, one should also consider the presence of specific molecules that ensure the production of a new cytoskeleton. The faint MAP 5 immunostaining, however, could also be caused by the presence of the MAP 5B protein (see Riederer 1990). Earlier, it was shown that the anti-MAP 5 antibody used in this study recognizes the MAP 5A and the MAP 5B protein (Riederer et al. 1986). The latter is still expressed in the adult (Riederer et al. 1991), whereas the (phosphorylated) MAP 5A protein is considered as a specific neuronal marker for the immature brain (Viereck et al. 1989). MAP 5A is present in extending processes and thought to be one of the regulators of axonal outgrowth (review by Tucker 1990). The use of antibodies specifically directed against only one of the two MAP 5 proteins could solve this confusion. In general, the overall decrease in MAP immunoreactivity is thought to be due to genetic downregulation, since it was demonstrated that MAP 5 mRNA decreases with development (Tucker et al. 1989).

The present results demonstrated an almost exclusive expression of MAP 2 in the spinal cord mantle layer, which is in agreement with earlier studies and shows the involvement of MAP 2 in the formation of dendritic processes (Bernhardt and Matus 1984; Caceras et al. 1984; De Camilli et al. 1984). Recently, it was demonstrated that mRNA for MAP 2 is also present in dendrites and could facilitate a rapid response to local demands (Garner et al. 1988). Although the exact roles of the three different MAP 2 isoforms are still obscure, it has been suggested that MAP 2A and 2B are associated with the formation of the dendritic processes as potent regulators of the attachment of microtubules to each other or to neurofilaments (Bernhardt and Matus 1984; review by Tucker 1990).

The expression of the MAP 2 antigen in the mantle layer followed a ventral-to-dorsal gradient. At E12, the antigen was present in the ventral region, 1 day later in the intermediate region, and around E14–E15 in the dorsal part of the mantle layer. This gradient clearly coincides with the developmental gradient which is found in the generation and migration of the spinal cord neurons (Altman and Bayer 1984). With regard to the ventral motor neurons, the present results are consistent with an earlier study demonstrating that the formation of the axon and dendrites of motor neurons starts at E12 (Altman and Bayer 1984).

One of the three isoforms of MAP 2, MAP 2C, is particularly abundant in the developing brain (Tucker et al. 1988). As well as in dendrites, this protein was also found to be expressed in axons and glial cells, which could suggest that MAP 2 staining in the developing mantle layer, as shown in this study, actually reflects a glial cell staining. However, its expression pattern does not correspond to the known glial cell patterns in the developing rat spinal cord (see below; Oudega and Marani 1991). Therefore, the idea that the MAP 2 immunostaining pattern in the mantle layer reflects its presence in glial extensions can be ruled out. Its presence in axons within the mantle layer, however, must be considered.

MAP 2 could still be detected in the mantle layer of adult spinal cord. Obviously, the dendrite-specific distribution of MAP 2 is established at the moment of dendritic outgrowth and is maintained throughout the maturation process. The capacity of the used antibody (clone C) to recognize three MAP 2 isoforms could actually result in a staining of the MAP 2B and MAP 2C isoforms during the early development of the cord and the detection of the MAP 2A isoform later on in development. The adult MAP 2 staining, as described in the present study, could therefore reflect the presence of the MAP 2A isoform.

With regard to the adult spinal cord, two special features in the MAP 2 expression patterns must be mentioned. First, MAP 2 was clearly present in Rexed's laminae I and II. In contrast, these laminae appeared to be almost devoid of MAP 1A. This may indicate that these two layers are particularly occupied by (MAP 2-positive) dendritic processes and are less rich in (MAP 2A-positive) axons. Second, MAP 2 was found to be present in processes that project into the marginal layer (see Sect. 5.3). Especially in the dorsal part of the lateral funiculus, a large number of MAP 2-positive fibers were detected. These processes lack the MAP 5 protein and, therefore, seem to represent dendrites of neurons present in the mantle layer.

The expression patterns of the studied cytoskeleton proteins, which can be regarded as structural morphoregulators, in the developing and adult spinal cord mantle layer seem somewhat predictable. All proteins followed the ventral-to-dorsal developmental gradient that is generally considered to be present in the development of the mantle layer. Most of them were found to demonstrate a transient expression, which indicates their involvement in developmental processes that determine the gross anatomy of the spinal cord. On the other hand, it seems unlikely that these proteins play prominent regulatory roles in particular developmental processes such as neuron migration, clustering, and differentiation. Their main role seems to be a supportive function in the formation and maintenance of a proper cytoskeleton.

The present results demonstrate SSEA-1 in the dorsal part of the mantle layer of the developing rat spinal cord. So far, SSEA-1 has not been demonstrated in the mantle layer, which has led to the suggestion that migratory or postmigratory neurons lack the antigen (Yamamoto et al. 1985). However, in this study the antigen was clearly discerned in several separate regions of the developing dorsal mantle layer from E17-E18 on. This finding suggests a role for SSEA-1 in migratory or postmigratory interneurons of the rat spinal cord mantle layer. Unfortunately, it could not be determined whether the neurons expressed SSEA-1 during or after their migration. Although difficult to determine, the three separate areas were identified in adjacent cresyl violet-stained sections as Rexed's laminae II, IV, and VI/ VII (including the lateral horn). During further development, the antigen disappeared from laminae IV and VI/ VII. Finally, in the postnatal and mature spinal cord, only lamina II remained positive for SSEA-1.

Based on these results, several functions for SSEA-1 in these particular regions could be predicted. First, the antigen could be involved in the regulation of the migration and settlement of the dorsal horn interneurons. Second, SSEA-1 could serve as a regulator in the establishment and maintenance of the contacts of C fibers and A(δ) fibers. Earlier, similar conclusions were drawn from an immunocytochemical study in which the mature rat spinal cord was found to contain several lactoseries carbohydrates (Dodd and Jessel 1985). The different possibilities will be discussed in detail in Chap. 6.

Soon after birth, faint, coarse SSEA-1 distribution was detected throughout the spinal cord mantle layer. The present study does not offer an explanation for this particular phenomenon, but one explanation could be the start of the myelination process, which is known to take place after birth. Earlier studies demonstrated that SSEA-1 can be found on sphingolipids (Urdal et al. 1983), a major component of the myelin sheath. This would actually imply that the used antibody (known as 3B9) is capable of recognizing its antigenic determinant when located on various molecular species. Antibodies detecting SSEA-1 in glycolipids before birth and in proteoglycans after birth were described earlier (Yamamoto et al. 1985). Conversely, it also shows that SSEA-1 most likely serves different functions during the development and in the adult rat spinal cord.

At E12, AChE was found to be expressed in the motor neurons of the incipient ventral horn. Between E14 and E16, the organization of these AChE-positive motor neurons into the cell columns appears to take place, which is in agreement with earlier studies (Angulo Y Gonzáles 1940; Altman and Bayer 1984). Taking this into consideration, together with the knowledge that the motor neuron processes are formed from E14 on (Altman and Bayer 1984), it seems unlikely that the AChE presence actually reflects cholinergic transmission. Rather, the demonstrated early expression of the enzyme in rat spinal cord motor cells may be associated with developmental processes such as the organization of the neuronal cytoarchitecture and/or the pathfinding of the neuronal processes. The involvement of AChE in such neuronal developmental events has been proposed before (Brown et al. 1986; Marani et al. 1986b; Hutchins and Casagrande 1988; Layer et al. 1988; Oudega and Marani 1990). Although roles of AChE other than its function as the breakdown enzyme of the transmitter ACh have been well documented over recent years, its precise role as a regulator of morphogenetic events in the central nervous system is still obscure (Kostovic and Rakic 1984; Layer and Sporns 1987). One of the more likely functions for AChE during development could be the facilitation of communication between neighboring cells, a role which is supported by the finding that AChE is capable of modulating neuronal cell membrane permeability (Rauschecker and Singer 1981; Sillito and Kemp 1983; Greenfield 1991). Such an action of the enzyme could lead to the coordinated synchronized development of the ventral motor neurons of the rat spinal cord.

During further development and at maturity, the motor neuron cell bodies remain positive for AChE, while their axons gradually appear to loose the enzyme. By now, the presence of AChE in the motor neurons could contribute to the stabilization of active synapses, a function that has been proposed for the enzyme during postnatal development of the central nervous system (Singer 1979). This particular action of AChE should be the result of its interference with transmission at excitatory synapses (Rauschecker and Singer 1981; Sillito and Kemp 1983). In functioning this way, AChE contributes largely to the final shaping of connections during critical periods in neuronal development. Such a developmentally important regulatory role for AChE has been demonstrated previously in the rat cortex (Bear et al. 1985). In some cases, the expression of a cholinergic phenotype within the spinal cord appears to be transient. Such a temporary cholinergic phenotype was also found to occur in neurons of the developing sympathetic system (for review see Landis and Reese 1983), which suggests a more widespread transient function for the enzyme.

In this study, the substantia gelatinosa (Rexed's lamina II), which is known to receive unmyelinated primary afferents (Réthelyi 1977), demonstrated an increase in AChE activity during the second postnatal week. This enhancement in the expression of AChE may indicate a "late" development of the nociceptive primary afferents. Traditionally, these fibers are known to convey the peripheral nociceptive information, in which several compounds were found to play an active role (Cuello et al. 1978). One of the best studied strutures is substance P (Hökfelt et al. 1976). With respect to the substance P presence, the enzyme expression may indicate an involvement of AChE in the hydrolyzation of this neuromodulator. Such a function of AChE has been proposed before (Chubb et al. 1980; Lockridge 1982).

The present results demonstrate an abundant expression of AChE in the intermediate gray from E15 on. A cluster of AChE-positive cells was found in the lateral aspect of the intermediate gray. Also, the area between this cell cluster and the central canal was found to express the enzyme. The laterally situated cell group represents the intermediolateral cell column, which is situated in the lateral horn of the adult rat spinal cord mantle layer. These cells are generated between E12 and E13, and the first cells settle in the future lateral horn around E14 (Altman and Bayer 1984). The area between this column and the central lumen is known as the intercalated region, which besides cells also contains extensions of the cells of the intermediolateral cell column (Forehand 1990). The intermediolateral cell column contains the sympathetic preganglionic motor neurons (Hancock and Peveto 1979b; Rando et al. 1981). Such neurons are also located in the so-called intermediomedial cell column and in the dorsal part of the lateral funiculus. The present study clearly demonstrated the presence of AChE-positive cells in the latter region, but whether these cells are actually sympathetic preganglionic motor neurons (see Sect. 5.3) remains to be established.

After P12, the AChE presence in the intercalated region diminished, whereas the cells of the intermediolateral cell column remained AChE positive until maturity. The present results correlate well with the cytoarchitectonic study by Altman and Bayer (1984) and seem to implicate that AChE is at least partially (namely, during the cell aggregation process) involved in the development of the autonomic system in the rat spinal cord. So far, the exact role of AChE in this kind of developmental process can only be speculative (see above).

The temporary expression of AChE in the projections of the autonomic neurons suggests a role in the development of these outgrowing fibers. Such a role was suggested earlier in this study in relation to the development of the ventral spinal motor neurons. The exact role of AChE as a fiber outgrowth-regulating substance is still unclear, but one possibility is that the enzyme is involved in the stabilization of active synapses by modulating the permeability of the cell membrane (Rauschecker and Singer 1981; Sillito and Kemp 1983). Unfortunately, data on the outgrowth of the autonomic projections and especially on the establishment of their peripheral synaptic contacts are still absent from the literature. Based on the results in this study, it is proposed that the development of the peripheral projections of the cells of the autonomic intermediolateral column starts around E18.

The overall vimentin- and GFAP-positive radial fiber pattern in the developing and mature mantle layer indicates the involvement of the two glial markers in the basic structural organization of the spinal cord (see also Aguayo et al. 1981; Silver et al. 1982). The radial orientation suggests a role for spinal cord glial cells in neuroblast migration. It was suggested earlier that the neuroblasts migrate orthogonally from the matrix layer towards their final location in the mantle layer (Altman and Bayer 1984). The demonstrated radial orientation of the vimentin/GFAP-positive glial fibers confirms such a direction of migration. Only recently, it was shown that migrating neuroblasts are actually associated with radially orientated glial processes (Gasser and Hatte 1990). Glial cells have already been found to serve as an appropriate scaffold for migrating neurons (Rakic 1972; Sidman and Rakic 1973). The time of appearance of the vimentin radial pattern in the mantle layer coincides remarkably well with the known times of neuroblasts migration in the rat spinal cord. Between E13 and E18, an abundant radial vimentin pattern is observed in the mantle layer, and it is known that most of the neuron migration takes place during this period (Altman and Bayer 1984). The appearance of the radial vimentin staining also demonstrates the ventral-to-dorsal gradient that was found to be present in neuronal migration in the spinal cord (Altman and Bayer 1984). From E13 on, radially orientated, vimentin-positive fibers were present in the intermediate gray region. This suggests a special involvement of the glial fibers in the migration of the spinal cords relay cells, which are known to be produced during a 4-day period, E11–E14 (peak production between E12–E13; see Altman and Bayer 1984).

At birth, GFAP fibers in the dorsal horn are orientated tangentially, which is in contrast to the general orientation of the spinal cord glial fibers. The glial fibers of the dorsal horn appear to converge towards the apex (or neck region) of the dorsal mantle layer. This pattern may indicate the involvement of

GFAP-positive glial fibers with ingrowing primary afferents. The latter are known to enter the dorsal horn by way of Lissauer's tract and to terminate in the upper laminae (Light and Perl 1979; Ralston 1979). The role of these particular GFAP-positive fibers could be the guidance of these ingrowing primary afferents towards their appropriate targets. The involvement of glial cells in the migration of neurons has recently been demonstrated (Gasser and Hatte 1990). Radially orientated glia cell processes are in close contact with migrating neurons in order to guide them to their final destination in the mantle layer (Gasser and Hatte 1990). The neurons place their cell body against glial fibers and extend a motile, leading process in the direction of the migration. This (neuronal) extension was found to enfold the glial fiber with short filipodia and lamellipodia (Gasser and Hatte 1990).

At E18, GFAP-positive fibers in the ventral horn were clearly associated with the motor neurons in the ventral motor columns. These GFAP-positive glial fibers did not originate from the matrix layer, as was established for the other GFAP-positive fibers in the mantle layer. Therefore, it is proposed that besides the generally radially orientated, GFAP-positive fibers of the mantle layer, a second local source of such fibers exists in the spinal cord. The precise function of these locally generated GFAP fibers remains to be elucidated, but a close association with the development of the spinal cord motor neurons seems likely.

Earlier, it was suggested that the resilience of glial fibers/cells in the roof plate of the matrix layer leads to the formation of the dorsal midline (Altman and Bayer 1984). Although the present study does not directly offer an alternative, it strongly suggests that the glial fibers which form the dorsal raphe originate in the intermediate part of the matrix layer. Day by day examination of the material indicates that from E15 on, vimentin-positive fibers are present in the dorsal raphe. During the second postnatal week, vimentin was found to be replaced by GFAP. The GFAP-positive fibers appeared to originate from the intermediate region of the matrix layer. The development of the ventral raphe showed large similarities with the dorsal midline. In the ventral raphe, a major vimentin barrier was also replaced by a thinner GFAP barrier after birth. The function of these raphe structures is widely debated, but one likely role of the dorsal raphe is to prevent growing axons from decussating to the contralateral half of the spinal cord.

5.3 Marginal Layer

The marginal layer of the rat spinal cord is occupied by different types of axons: supraspinal, descending projections from the brain and ascending axons towards the brain course along the ventral, lateral, and dorsal funiculi. The fibers that connect different levels of the spinal cord, the so-called propriospinal axons, run in the ventral and lateral funiculi close to the mantle layer. The dorsolateral funiculus contains sensory fibers that project towards the thalamic region of the brain.

The ipsilaterally and contralaterally projecting neurons of the intermediate mantle layer project their axons into the lateral and ventral funiculus, respectively. At E13, the first of these fibers can already be detected in both funiculi (Altman and Bayer 1984; Oudega et al. 1990a). Recently, it was demonstrated that the spinal commissural axons are guided in their decussation by the cells of the floor plate (Tessier-Lavigne et al. 1988; Placzek et al. 1990). At E13, the incoming dorsal root fibers enter the cord at the dorsal root entrance zone and split into an ascending and descending branch in the oval-shaped dorsal root bifurcation zone. These structures shift medially towards the dorsal midline of the cord and form the dorsal funiculus (Altman and Beyer 1984; Oudega et al. 1990a). A more thorough description of this developmental phenomenon will be provided later on in this chapter. Most of the dorsal root fibers (primary afferents) enter the cord via the dorsal roots, although some were found to enter via the ventral roots (Coggeshall et al. 1974). The populations of large and small axons that enter the cord were found to be segmented in the dorsal roots (Light and Perl 1979). Earlier, it was demonstrated that around E16, some of the dorsal root fibers bundle after entering the spinal cord and course towards the ventral horn, which they reach 1 day later (Kudo and Yamada 1987). Most likely, these particular primary afferents are part of the spinal reflex arc.

The first descending supraspinal fibers can be distinguished in the ventral funiculus and belong to the reticulospinal tract (Ramon Y Cajal 1909). Several studies using the neuronal tracer DiI have showed that the supraspinal control on spinal cord neurons develops soon after the closure of the neural tube. For instance, the first fibers from the interstitial nucleus of Cajal reach the cervical levels of the spinal cord around E12. At E14, axons from the vestibular complex appear to reach the lower cervical levels of cord. The supraspinal serotoninergic innervation of the spinal cord has been studied extensively (see Rajaofetra et al. 1989). Around E14, serotoninergic projections from the raphe nuclei invade the spinal cord and 2–3 days later reach its caudal levels. Around E15, the first serotoninergic fibers appear to invade the cervical gray matter (Rajaofetra et al. 1989). On the same day, descending, supraspinal, noradrenergic fibers were found in the lateral funiculus (Bernstein-Goral and Bohn 1988). From these studies, it appears that the time period E13–E16 is critical in the development of the descending supraspinal projections.

During recent decades, spinal cord fiber systems have been extensively studied by means of neuronal tracers (Arsénio-Nunes and Sotelo 1985; Gribnau et al. 1986; Joosten et al. 1987; Lakke and Hinderink 1989; Lakke and Marani 1991). So far, the studies of the embryonal development of the different fiber systems have been limited, due to technical reasons. In the present study, however, the development of several supraspinal fiber tracts was studied with an intrauterine tracing technique. These fiber systems will be discussed separately.

After birth, the myelination of the axons in the spinal cord takes place. This process continues during the first 3 postnatal weeks and results in the characteristic white opaque appearance of the spinal cord funiculi. The oligodendrocytes, which derive from the O2A progenitor, regulate the myelination of the axons. It has been found that neurons themselves modulate the myelin protein mRNA levels in oligodendrocytes (Kidd et al. 1990). In other words, neurons influence their own myelination and therefore improve their performance as signal conductors.

The outgrowth of neuronal processes in general and the extension of one in particular is considered to be one of the most intriguing processes in developmental neurobiology. A large group of different factors seem to influence and/or regulate not only the outgrowth and extension of the neurites, but also their guidance towards the target (for a recent review on growth factors with influence on axon elongation, see Lipton 1989).

Axons appear to select specific pathways by recognizing special cues, e.g., cell surface and extracellular matrix molecules that mediate cell and substrate adhesion and fasciculation. These extracellular molecules function as attractants which the developing axons use as a substrate for growth (Kroger and Walter 1991). It was demonstrated that the support and guidance of growing axons is also regulated by repulsion and inhibition (see Walter et al. 1990). Recently, it was shown that some adhesion molecules serve as permissive substrates; they are able to define the pathway, but they do not provide information about which path to take where there is a choice (Lemmon et al. 1992). The growth cones were found to be the critical structure in axonal pathfinding (Tosney and Landmesser 1985; Bentley and Caudy 1983). These growth cones are able to recognize and discriminate between different molecules, and by doing so they guide the neurite towards the proper target (Kuwada and Bernhardt 1990). Diffused molecules play a crucial role in finding the target and establishing the connections in the brain (Heffner et al. 1990). After having reached their target, the axons develop synaptic contacts, which either become active or will be removed (remodeling or rearrangement of the connections). In order to survive, the axon actually needs to establish a synaptic connection with the intended target cell. As well as the axon, the outgrowth of dendrites was also found to be dependent on neuronal adhesion (Chamak and Prochiantz 1989). A remarkable difference between the development of the axons and the dendrites is the high speed at which axons appear to elongate, a difference that most likely is reflected in the molecular composition of their cytoskeleton. In the adult, the cooperation between the contact-dependent compounds and the diffusible growth factors during development finally, results in the formation of the central nervous system with its complex pattern of axonal and dendritic connections.

Among the neurite outgrowth promoting factors is fibroblast growth factor (Morrison et al. 1986; Unsicker et al. 1986; Walicke et al. 1986). Recently, the involvement of the so-called 2A1 antigen in the extension of neurites has been demonstrated (Schlosshauer et al. 1990). Other modulators of neurite outgrowth are the Thy-1 receptor, which was found to be present on glial cells (Dreyer et al. 1990), spectrin (Barakat-Walter and Riederer 1991), and GAP-43 (growth-associated protein-43), which is predominantly found in growth cones and nerve terminals (Skene et al. 1986; Goslin et al. 1988). A direct involvement of the latter protein has been suggested, but so far conclusive data have not been obtained.

Extracellular matrix molecules are known to function as a substratum for growing axons (Kroger and Walter 1991). Pathways for neurite guidance mediated by adhesive forces may actually be formed in the extracellular matrix by means of laminin (Lander et al. 1985), cytotactin (Grumet et al.

1984), or myelin-associated glycoproteins (Poltorak et al. 1987). Other known guidance-mediating compounds are glycoproteins (Jessel 1988), fibronectin (Horwitz et al. 1986), TAG-1 (Dodd and Jessel 1988), and N-CAM (Edelman 1983).

The growth of an axon was found to be stopped either by a physical barrier or by a particular chemical factor. In vitro experiments have shown that oligodendrocytes are able to produce nonpermissive substrate properties which prevent neurite growth (Wictorin et al. 1990). Savio and Schwab (1990) showed the existence of two inhibitory proteins with a molecular weight of 35 and 250 kDa on oligodendrocytes and in the myelin they produce. Recently, it was demonstrated that these inhibitors appear in the spinal cord just after the period of axonal growth and before the start of the myelination process (Caroni and Schwab 1989).

In general, the studied protein members of the axonal cytoskeleton display a largely similar distribution pattern in the developing ventral and lateral funiculus. The exception to this rule was the distribution pattern of the MAP 2 protein, which does not occur at all within the fibers of the spinal cord funiculi. Nevertheless, MAP 2 could be detected in the funiculi, but as a component of processes originating in the mantle layer (see Sect. 5.2). This absence of MAP 2 in the marginal layer fibers is not surprising in view of the fact that this molecule is known as a MAP associated with dendrites. Of the three MAP 2 isoforms, however, the MAP 2C molecule has also been detected in axons. Obviously, the present antibody, which is capable of recognizing all three isoforms, is not able to distinguish its antigen in the long ascending and descending fiber systems of the rat spinal cord.

Before discussing the expression patterns of the neurofilaments in the developing marginal layer of the rat spinal cord, the specificity of the used antibody and some of its functional implications should be mentioned. MAb NF-90 is capable of recognizing the phosphorylated forms of the three neuro-filament subunits (Oudega et al. 1990a). Phosphorylation usually takes place after the subunits have entered the axon and results in the cross-linking and stabilization of the subunits into neurofilaments (Hirokawa et al. 1984; Carden et al. 1985; Georges et al. 1986; Nixon et al. 1987; Oblinger 1987). Besides being a prerequisite for axonal spacing (see Matus 1988), phosphorylation of the subunits als provides a mechanism for interaction between neurofilaments and other cytoskeletal elements and stabilization of the axonal cytoskeleton (Hirokawa et al. 1984; Nixon et al. 1987; Dahl 1988). Obviously, the process of phosphorylation plays a key role in the remodeling of the neuronal cytoskeleton, which is of critical importance for neuron growth, axon guidance, and signal transduction. With regard to this important developmental role, the two smaller (phosphorylated) NF-L and NF-M subunits seem especially involved, as they were found to be present in the immature axon. The phosphorylation of the NF-L subunits is thought to be a means of preventing its early assembly and/or of disassembling already formed filaments (see Nixon and Sihag 1992), which seems to be of importance during the early stage of axonal extension. The largest of the three subunits (NF-H) was found to be expressed later on in the maturation process.

At E13, the ventral funiculus and the ventral commissure demonstrate the first neurofilament-positive axons. During the next few days, a concentration of neurofilament-positive fibers was discerned in the periphery of the ventral funiculus (see below). These fibers might represent the first ascending axons of the contralaterally projecting neurons of the intermediate gray. However, considerable contribution by descending supraspinal fibers cannot be excluded, as these fiber systems were found in the spinal cord around E14–E16 (see Rajaofetra et al. 1989). Among these descending fiber systems, axons of the reticulospinal tract are also to be expected, as careful examination of E14–E15-aged brains revealed that this tract had already largely developed. In the chick, the first descending fibers to be recognized in the ventral funiculus also belong to the reticulospinal tract, a component of the fasciculus longitudinalis medialis (Ramon Y Cajal 1909; Windle and Austin 1936; Okado and Oppenheim 1985). It should be mentioned that these assumptions are based on the knowledge that the ontogeny of the descending supraspinal systems in the chicken is similar to that seen in a wide variety of vertebrates (Ten Donkelaar 1982; Martin et al. 1978, 1982). From P12 on, the subsurface region of the ventral funiculus gradually developed a concentration of neurofilament staining. This phenomenon will be discussed below.

At E12, the first ascending axons of the ipsilaterally projecting neurons of the intermediate gray are present in the lateral funiculus. Consequently, the ipsilaterally projecting fibers develop ahead of the contralaterally projecting fibers, which course along the ventral funiculus. In the chick, the earliest fibers in the spinal cord are also located in the incipient lateral funiculus (Ramon Y Cajal 1909; Bok 1928; Windle and Austin 1936; Nornes et al. 1980a,b; Okado and Oppenheim 1985). The lateral funiculus of the mature spinal cord is known to contain long ascending and descending supraspinal fiber tracts as well as propriospinal fibers (Giesler et al. 1981; Cabana and Martin 1982; Martin et al. 1983; Arsénio-Nunes and Sotelo 1985). Obviously, it cannot be concluded from the present results that the first neurofilament-positive fibers belong to a particular fiber tract. Most likely, all three different fiber systems participate in the first contingent of neurofilament-positive fibers, whereas in the chick spinal cord it appears that these systems develop almost simultaneously (Okado and Oppenheim 1985). Besides matching these data, the present neurofilament results are also in agreement with the early development of spinal cord fiber systems of the North American opossum (Martin et al. 1983).

Until E16, it seems that the periphery of the lateral funiculus contained a higher number of fibers. Between E18 and P4, more (immunoreactive) fibers were found in the subsurface area of the lateral funiculus, just under the most peripheral region. The ventral funiculus also displayed a peripheral concentration of neurofilament-positive fibers during early development (see above). The first fibers to develop possibly gather in the periphery. Later on, in the lateral funiculus these early fibers are packed more deeply in the region, because the later-formed axons grow along the surface of the cord. Such a developmental event has been demonstrated before in the chicken, in which the (later-formed) spinocerebellar fibers are located in the periphery of the spinal cord (Nornes et al. 1980a; Okado and Oppenheim 1985; Okado et al.

1987). From a phylogenetic point of view, this development process is highly intriguing. The earliest fibers that develop, first located in the periphery and later on in the deeper region of the cord funiculi, innervate the phylogenetically older brain regions (i.e., thalamus). In the phylogenetically younger regions (i.e., cerebellum), the efferent spinal fibers develop at a later stage and course along the outer, more superficial region of the funiculi. In doing so, these younger fibers push the earlier-formed fibers to the deeper areas.

After P12, the subsurface region of the lateral funiculus and the peripheral ventral funiculus still exhibited a more intense neurofilament staining. By now, it appears that a larger amount of large caliber fibers is actually present in these areas. The neurofilament-rich areas in both funiculi were found to be continuous with each other. Probably, most of the thick axons belong to the fasciculus longitudinalis medialis, which is known to contain a large contingent of large caliber fibers.

The most dorsal part of the lateral funiculus contained only a few neurofilament-immunoreactive fibers at the early embryonic ages. During later embryonal and postnatal development, first the area just beneath the dorsal horn and later on the deeper region fills with neurofilament-positive fibers. In the deeper region, fibers first seem to occupy the more medial part. These results may indicate that the first fibers of the rubrospinal tract, which is located in the deeper region of the dorsolateral funiculus (Waldron and Gwyn 1969; Brown 1974b), reach the cervical cord around E18 and the lumbar cord around E20. The bulk of rubrospinal fibers arrives relatively late and according to a rostral-to-caudal gradient. Around P4–P6, fibers were found in the cervical part and around P8–P10 in the lumbar part of the spinal cord, which suggests an extended period of development for the rubrospinal tract. These findings are largely in agreement with results obtained after intrauterine use of WGA-HRP in the rat spinal cord, which confirmed the presence of the first rubrospinal fibers at cervical levels at E17 (Lakke and Marani 1991). These results will be extensively discussed in the second part of this chapter. It must be kept in mind that the appearance of the phosphorylated neurofilaments in the dorsolateral funiculus could also reflect the maturation of the rubrospinal fibers, as phosphorylation of neurofilaments is closely related to axonal maturation (Dahl and Bignami 1986; Bignami and Dahl 1987; Dahl 1987). Therefore, it could be that the rubrospinal fibers are actually present, but not detectable with the neurofilament antibody used in this study. However, such a phenomenon seems unlikely, as it was demonstrated earlier that with the Bodian fiber staining, axons could be demonstrated at the same time intervals as described above with neurofilament immunocytochemistry (see Oudega et al. 1990a). The Bodian staining technique recognizes the phosphorylated as well as the nonphosphorylated neurofilament subunits (Oudega et al. 1992a).

From E13 on, the fibers in the different funiculi of the rat spinal cord express MAP 1A and, especially, MAP 5. The unexpected presence of the MAP 1A antigen during these early days of spinal cord development is discussed above. Later in development, both proteins were found to vanish from the spinal cord funiculi. The temporary presence of both proteins in the marginal layer suggests their involvement in the development of the various fiber systems of the funiculi of the rat spinal cord, i.e., the long ascending and descending supraspinal fibers and the propriospinal fibers. Both proteins are known for their regulatory role In the assembly and stabilization of microtubules (Kuznetsov et al. 1981; Riederer et al. 1986). The MAP 5 protein was abundantly expressed in the funiculi of the developing spinal cord. Earlier, it was demonstrated the MAP 5 is capable of stimulating tubulin polymerization in vitro (Riederer et al. 1986). Therefore, it seems likely that MAP 5 is involved in the active growth of neuronal processes. During axon elongation, the continuing growth of the microtubules does require the presence of MAP 5 in order to assemble the tubulin subunits. Beside the assembly and stabilization of microtubules, MAP 5 might also be involved in the cross-linking of micro-tubules with other filaments of the axonal cytoskeleton, which results in the increase of the axonal diameter. With regard to this possibility, it is of interest that the adult microtubular density was found to be established soon after birth, when the radial growth of the axons is still taking place (Faúndez and Alvarez 1986).

After P12, MAP 1A-positive cells were found scattered throughout the funiculi of the rat spinal cord. These small, round cells most likely represent glial cells. The presence of MAP 1A (and MAP 1B) within glial cells was demonstrated earlier (Bloom et al. 1984). On the other hand, Huber and Matus (1984) could not demonstrate MAP 1A in glial cells with a monoclonal antibody similar to the one used in the present study. These immunocyto-chemical differences might reflect the differences in the fixation of the nervous tissue, as was explained and discussed before (see also Oudega 1990; Oudega et al. 1990a). MAP 5 could not be detected within glial cells in our study, which is in agreement with earlier reports on other parts of the central nervous system (Riederer et al. 1986; Tucker et al. 1988). The function of MAP 1A in glial cells is still obscure. Whether the MAP 1A protein has a similar role as in the axonal cytoskeleton or a totally different function remains to be solved.

In the marginal layer, the intensity of MAP 1A and MAP 5 staining decreased after the first postnatal week. This may reflect the switch from a growing phase to a period of stabilization and maintenance of the axonal cytoskeleton. During the same period, a similar change has been demonstrated in other parts of the rat brain (Riederer 1986; see review by Matus 1988). Obviously, the first 3 weeks after birth are of vital importance within the central nervous system with regard to the change from a juvenile (developing) state to a mature (maintenance) state. The faint MAP 1A and MAP 5 immuno-reactivity that seems to remain present in the mature spinal cord funiculi may reflect the presence of small amounts of these proteins in the adult axons. This might demonstrate a constantly low turnover rate of these molecules, which could benefit a certain degree of plasticity in the axons, as discussed above.

During development, SSEA-1 could not be associated with the fibers of the developing ventral and lateral spinal cord funiculi. However, a small region dorsally located in the lateral funiculus exhibited SSEA-1 during a short period, E13–E14. The immunoreactivity did not appear to be linked with the present fibers. A possible role for SSEA-1 could be the guidance of incoming primary afferents towards the dorsal root bifurcation zone. This would stress again the association of SSEA-1 with the rat sensory fibers. After birth, a coarse SSEA-1 staining pattern gradually developed. The most likely explanation for this finding would be the myelination process, which takes place during the first few postnatal weeks. Previous studies have demonstrated that SSEA-1 can be found on sphingolipids (Urdal et al. 1983), a major component of the myelin sheath. Also, antibodies detecting SSEA-1 in glycolipids before birth and in proteoglycans after birth have been described (Yamamoto ct al. 1985). Consequently, it seems likely that the antibody used in the present study, designated 3B9, recognizes its epitope when present on different molecules: before birth, on glycolipids present in the neuronal membrane, and after birth in proteoglycans in the myelin sheath. However, the present study does not supply clear evidence for the presence of SSEA-1 in the myelin sheaths of postnatal and mature spinal cord axons. One approach towards solving this question could be the demyelination of the spinal cord sections before their immunocytochemical staining. This should result in negative staining by the loss of the antigen.

The function of the SSEA-1 molecule as part of myelin sphingolipids is unclear. One speculative explanation, at least for its presence during the later developmental stages, may be that it continues its role in axon guidance. If this is the case, the myelin sheath not only acts as a permissive substrate (see Savio and Schwab 1990; Wictorin et al. 1990), but also functions as a guiding/attracting substrate for growing axons. Such an involvement of myelin-associated extracellular matrix molecules in the guidance of neurites has been suggested previously (Poltorak et al. 1987).

In the present study, it was shown that a cluster of AChE-positive cells is present in the dorsal aspect of the lateral funiculus. These cells were detected between E16 and E21 and are present at all spinal cord levels, though most abundantly at the thoracic level. Previously, the dorsal aspect of the thoracic lateral funiculus was found to contain sympathetic, preganglionic neurons (Petras and Cummings 1972; Hancock and Peveto 1979a,b; Rando et al. 1981). Cells of the lateral spinal nucleus settle in the same region, but unlike the sympathetic neurons, they can also the detected at the cervical and lumbar levels (Gwyn and Waldron 1968). At least part of the AChE-positive cells, as demonstrated in this study, belong to the lateral spinal nucleus. Obviously, at the thoracic level, the two different kind of cells are intermingled and, therefore, difficult to distinguish from each other. It is possible that both cell types contain the enzyme, The exact role of AChE in these cells is still unclear and remains to be solved in future research. Due to its temporary presence, it seems most likely that the enzyme functions in their development.

As early as E14, nicely arranged palisades of vimentin-positive fibers were demonstrated in the ventral and lateral funiculus. Such an arrangement of vimentin was detected until the first postnatal week. Starting at the end of this week, a transition from vimentin expression to GFAP expression was found to occur in the developing funiculi. Nevertheless, as early as E18, the first GFAP-positive fibers were present in the ventral funiculus. Apparently, these fibers originate from a local source in the ventrolateral region of the ventral horn (see above). Later on, in addition to these fibers, another set of GFAP-position fibers were also demonstrated in the ventral and lateral funiculi. These glial fibers seemed to originate in the matrix layer. Earlier, a function of glia in the control of fiber growth was proposed, in which the glial fibers form a scaffold

116

for growing axons. In other words, they form a structural matrix for developing fibers (Singer et al. 1979; Silver et al. 1982; Hankin and Silver 1986). The present results clearly demonstrated an orderly structural configuration of glial fibers within the marginal layer of the spinal cord, which may indicate such a role of glia in the development of the descending and ascending fiber systems.

The dorsolateral fasciculus of the mature rat spinal cord contains thin ascending and descending fibers of the dorsal ganglion cells (Brodal 1981). These fibers terminate either in the dorsal horn at more rostral or caudal levels or traverse towards higher brain regions such as the thalamus. At E16, the first neurofilament-positive fibers in the development dorsolateral fasciculus were demonstrated. At the same age, the MAP 1A and MAP 5 proteins were sparsely present in the fibers of the dorsolateral fasciculus. Based on these results, it seems likely that the development of the fibers of the dorsal ganglion cells that course along the dorsolateral fasciculus starts around E16.

The two studied functional morphoregulators SSEA-1 and AChE did not appear in the dorsolateral fasciculus and apparently do not play a role in the development of its fibers. Vimentin and GFAP were only faintly present, which may suggest a less prominent role of these structures in the developing and mature dorsolateral fasciculus.

As compared to the ventral and lateral funiculi, the dorsal funiculus demonstated a more peculiar development. For this reason, the following section on the expression of the studied markers in the dorsal funiculus starts off with a short description of its development.

Around E13, at the dorsolateral aspect of the neural tube, primary afferents assembled in the dorsal roots enter the spinal cord at the so-called dorsal root entrance zone. The fibers form the oval-shaped dorsal root bifurcation zone, in which they split into an ascending and a descending branch (see Altman and Bayer 1984; Oudega et al. 1990a). It is believed that these fibers take care of the spinal cord intrasegmental connections, as at this age the dorsal root bifurcation zones are not fully continuous with one another (see Altman and Bayer 1984). Between E13 and E16, the dorsal root bifurcation zone develops towards the dorsal midline. The oval-shaped dorsal root bifurcation zone is now spread over the dorsal horn and is thought to contain the intersegmental fibers (Altman and Bayer 1984). After E17, the dorsal funiculus grows downwards, and a fasciculus gracilis and fasciculus cuneatus, which contain the suprasegmental, ascending primary afferents, can be recognized (Altman and Bayer 1984; Oudega et al. 1990a). Although it is generally accepted that large, myelinated fibers ascend in the dorsal funiculus, a small portion of unmyelinated axons can also be found in the dorsal columns (Patterson et al. 1989). The ventral aspect of the dorsal funiculus contains the rat corticospinal tract (Brown 1971; Gribnau et al. 1986). In the adult spinal cord, collaterals from the descending branch of the incoming primary afferents course ventrally along the medial dorsal horn, curve dorsally, and terminate in the nucleus proprius or Rexed's laminae III and IV (Light and Perl 1979). Other collaterals of the thick afferents terminate in the intermediate gray region (Lamotte 1977) and in the ventrally located motor columns (Brown 1981). Collaterals of the descending branch of the thinner primary afferents enter the dorsal horn dorsally and terminate in the marginal layer (Rexed's lamina I) or the substantia gelatinosa (Rexed's lamina II; Réthelyi 1977; Light and Perl 1979). The ascending collateral of the large-diameter primary afferents join in the dorsal funiculus. Those entering the cord at the sacral, lumbar, or low thoracic level gather in the medially situated fasciculus gracilis, and those from the higher thoracic and cervical levels assemble in the laterally situated fasciculus cuneatus (Walker and Weaver 1942). These ascending fibers terminate in the nucleus gracilis and cuneatus (Walker and Weaver 1942; Sterling and Kuypers 1967).

At E13, neurofilament-positive fibers could be detected in the dorsal root entrance zone, demonstrating the early primary afferents penetrating the cord. The dorsal root bifurcation zone also contained neurofilament-positive fibers, which seemed to be packed in its dorsal tip. These fibers are thought to provide the intrasegmental connections (Altman and Bayer 1984). At E16, a concentration of longitudinally coursing fibers was found in the most medial part of the initial dorsal funiculus, close to the dorsal midline. Two days later, such a concentration of positive fibers appeared to be present in the most dorsal area of the fasciculus gracilis. It must be noted that the present study did not clarify whether the concentration of neurofilaments was due to a higher number of fibers or to the presence of more (phosphorylated) neurofilaments per fiber. The amount of phosphorylated neurofilament subunits reflects the maturation of axons (Dahl and Bignami 1986; Bignami and Dahl 1987; Dahl 1987). Previously, ascending sensory fibers in the fasciculus gracilis were antegradely traced until the mid-cervical level (Wessels 1991). The same study demonstrated that after injections in the hindlimb, gracilis fibers reached the nucleus gracilis around E18–E19. Therefore, it seems plausible that the neurofilament-positive fibers in the fasciculus gracilis at the lumbar and thoracic levels, as demonstrated in this study, do indeed represent the ascending sensory fibers that enter the cord at the lower levels. So far, it is not known whether the primary afferents enter the cord according to a specific developmental gradient. The possibility exists that the ascending fibers of the lower levels course along the fasciculus gracilis and are pushed to the most dorsal tip by the newly entered axons, which would imply a caudal-to-rostral gradient in the development of the dorsal funiculus. However, this is contradictory to the earlier recognized rostral-to-caudal gradient, which is also the general developmental gradient in the rat spinal cord (Windle and Baxter 1936).

From E18 on, the lateral aspect of the fasciculus cuneatus showed more neurofilaments, which reflected a larger amount of thick dorsal root fibers. Within the fasciculus cuneatus, a gradually increasing amount of thick fibers was demonstrated during development. These fibers are the suprasegmental ascending collaterals of the thicker primary afferents that enter the cord at the higher thoracic and cervical levels (Walker and Weaver 1942; Altman and Bayer 1984). It was also noticed that some of the primary afferents penetrated the dorsal horn to terminate in Rexed's lamina IV. These fibers most likely represent collaterals from the descending branch of the incoming thicker primary afferents (Light and Perl 1979). They probably contact relay cells of the spinal cord intermediate gray.

Based on the expression of phosphorylated neurofilaments, the development of the corticospinal tract, which courses along the most ventral part of the dorsal columns, coincides remarkably well with the developmental patterns described earlier. The expression of neurofilaments in this area followed a clear rostral-to-caudal gradient. At E18, ventrally in the dorsal funiculus, a few neurofilament-positive axons appear to be present. Most likely, these fibers are the "pioneer" fibers described by Schreyer and Jones (1982). Later on, the bulk of positive fibers was found at the cervical levels at E20–E21 and reached the lower lumbar regions during the second postnatal week. Neuronal tracer studies revealed the presence of corticospinal fibers at the upper cervical segments on the first postnatal day (Schreyer and Jones 1982; Kort et al. 1985; Gribnau et al. 1986), whereas the lower segments were innervated during the second postnatal week (Gribnau et al. 1986; Joosten et al. 1987). The present results, therefore, confirm such an extended development of the corticospinal tract in the rat spinal cord. As stated before, the appearance of the phosphorylated proteins in the ventral part of the dorsal funiculus may also indicate the maturation of the corticospinal fibers, as phosphorylation of neurofilaments increases with axonal maturation (Dahl and Bignami 1986; Bignami and Dahl 1987; Dahl 1987). However, it was recently demonstrated that the corticospinal fibers could not be detected earlier in development when using Bodian silver staining, a technique that recognizes the phosphorylated as well as the non-phosphorylated neurofilament subunits (Oudega et al. 1990a).

From E13 on, MAP 1A and MAP 5 were expressed in the (intrasegmental) fibers of the developing dorsal funiculus of the rat spinal cord. After the development of the fasciculus gracilis and cuneatus (around E18), MAP 1A appeared to be lacking in the fasciculus gracilis, whereas the (suprasegmental) fibers of the fasciculus cuneatus abundantly expressed this particular protein. MAP 5 was abundantly expressed in both fasciculi during the embryonal period. The unexpected presence of MAP 1A during the early days of development has already been discussed above, as well as the expected abundant appearance of the MAP 5 antigen. The findings indicate that the suprasegmental, ascending primary afferents which enter the cord at the lower levels and which are actually present in the fasciculus gracilis, as demonstrated earlier (Wessels 1991), need MAP 5, but do not require MAP 1A, for proper organization of their microtubules during development. Surprisingly, however, those that enter the spinal cord at higher levels do require MAP 1A and MAP 5 to organize their microtubules. The precise reason for this lack of MAP 1A in the fasciculus gracilis fibers is unclear. Most likely, the proteins, where present, function as modulators of the assembly and stabilization of the microtubules in the growing primary afferents.

In contrast to the other parts of the white matter, until the second postnatal week the most ventral part of the dorsal funiculus, which contains the corticospinal tract, was found to lack the MAP 1A protein. MAP 5 could be detected in this area during development and at maturity. After P8–P10, however, MAP 1A appeared to be abundantly present in the ventrodorsal funiculus, which is in agreement with earlier studies (Bloom et al. 1984). These results indicate that in contrast to MAP 5, the descending fibers of the corticospinal tract do not require MAP 1A for their proper development. This difference between the corticospinal system and the other supraspinal fiber tracts, which do seem to need MAP 1A for their proper development, is surprising and needs to be clarified in future research. The assembly and stabilization of the axonal microtubules of the descending corticospinal fibers seem to be regulated solely by the MAP 5 protein. MAP 1A, however, is active in the mature phase of the corticospinal fibers, a role that was actually recognized and described earlier (Bernhardt et al. 1985; Riederer and Matus 1985).

After P12, MAP 1A-positive glial cells were detected in the dorsal funiculus of the rat spinal cord. The presence of MAP 1A (and MAP 1B) within glial cells was demonstrated earlier (Bloom et al. 1984; Oudega et al. 1990a,b; the present study) and discussed above. MAP 5 could not be detected within glial cells in our study, which is in agreement with earlier studies on other areas in the central nervous system (Riederer et al. 1986; Tucker et al. 1988).

During a short period (E13–E14), SSEA-1 appeared to be present in the dorsal root entrance zone. Through cell–cell contacts, carbohydrates are likely to play a role in guiding processes. Therefore, it seems plausible that the incoming primary afferents are guided by SSEA-1 towards the bifurcation zone. This again demonstrates a special relationship between SSEA-1 and the sensory afferents.

Soon after birth, the dorsal funiculus showed an overall distribution of SSEA-1. Although weak in intensity, the funiculus displays a coarse staining pattern. A possible explanation for this finding would be the start of the myelination process, which is known to proceed during the first 2 postnatal weeks. Earlier studies demonstrated that SSEA-1 is also found on sphingolipids (Urdal et al. 1983), a major component of the myelin sheath (see above).

The present study demonstrated the abundant presence of AChE in the dorsal root entrance zone and in the dorsal root bifurcation zone, which indicates in involvement of the enzyme in the early development of the centrally projecting (intrasegmental) fibers of the dorsal ganglion cells. The enzyme appeared not to be located within the fibers themselves, but seemed present throughout the dorsal funiculus. A possible role for AChE could be the guidance of the ingrowing fibers, whereas a function as terminator of the action of the neurotransmitter acetylcholine (ACh) seems unlikely at this stage development. Later on in development, AChE appeared to diminish from both fasciculi in the dorsal funiculus. Such a transient expression and the fact that AChE is located outside the fibers underlines the earlier proposed and above-mentioned role of AChE in developmental processes in the guidance of growing fibers. Earlier, it was demonstrated that the ascending suprasegmental fibers of the fasciculus gracilis are likely to be guided by AChE (Wessels 1991). At E17, these particular fibers, which originate from the dorsal root ganglion cells that have their peripheral process innervating the hindlimb, could be antegradely traced in the dorsal tip of the fasciculus gracilis until the high thoracic levels. At this level, the fasciculus gracilis appeared to contain high amounts of AChE (Wessels 1991). The tip of the fasciculus gracilis at the cervical level, which was not yet reached by the fibers, lacked the enzyme. Previous studies also indicated such a functional link for the enzyme in the chicken and stated AChE as a prerequisite for fiber growth and pathfinding (Layer et al. 1988). During development, the ventral part of the dorsal funiculus, which is known to contain the corticospinal tract (Gribnau et al. 1986), lacked the enzyme. In contrast to the fibers of the dorsal funiculus, the descending corticospinal tract fibers apparently do not need AChE for their proper development.

In the dorsal funiculus, a radial arrangement of vimentin-positive fibers was clearly demonstrated. Around E14, this region demonstrated a similar palisade pattern to the one found in the other funiculi of the rat spinal cord (see above). This suggests an involvement of glial fibers in the early development of the dorsal funiculus fibers. During the first few postnatal weeks, the transition from vimentin expression to GFAP expression within the dorsal funiculus was demonstrated. Earlier, Joosten and Gribnau (1989) also showed such a transition from vimentin to GFAP expression in the dorsal columns during roughly the same time period. The present study demonstrates that during this time frame, the vimentin–GFAP transition also takes place in the other parts of the marginal layer (see above). This coordinated transition throughout the spinal cord marginal layer suggests a genetically regulated time switch in the protein production of the glial cells. A possible initiator of this transition could be the fulfilment of the vimentin-regulated developmental processes.

Earlier, a function of glial cells in the control of fiber growth was proposed, whereby glial fibers acted as a structural matrix for axonal growth (Singer et al. 1979; Silver et al. 1982; Hankin and Silver 1986). The present results clearly demonstrated an orderly, structural configuration of glial fibers within the dorsal funiculus, which may indeed indicate a scaffold role for glial cells in the development of the ascending sensory and the descending corticospinal fiber systems.

The present study demonstrated longitudinally arranged palisades of vimentin-position glial cells in the dorsal funiculus. This distribution pattern has been described before (Joosten and Gribnau 1989). Based on electron microscopy experiments, the presence of adhesive type of contacts between the glial cells and the descending corticospinal axons was suggested (Joosten and Gribnau 1989). As well as a positional role, a chemical influence in the guidance of the corticospinal axons was also proposed. An additional function for the glial cell presence in the dorsal funiculus could be to prevent the decussation of the fibers. This possibility is underlined by the fact that such a glial cell septum appears to be absent in the originally decussating area of the pyramidal tract.

5.4 Development of Long Descending Systems

5.4.1 Methodological Considerations

Endocytosis of tracer macromolecules such as HRP and WGA–HRP occurs throughout the neuron membrane, including the membrane of axons and their terminals (Mesulam 1982). Endocytosis is rare along the myelinated part of the axon (Holzman and Peterson 1969; LaVail and LaVail 1974). Since myelination in the spinal cord only starts postnatally (Rozeik and Von Keyserlingk 1987), (WGA–)HRP can be taken up into all fibers that reach into, or course through, injection sites in the prenatal spinal cord. This notion is corroborated by the fact that in series in which the injection site had spread through both sides of the spinal cord, retrograde labeling (e.g., of the NR) was bilaterally symmetrical, even though the needle track (and damage) was located unilaterally in the spinal cord. This indicates that endocytosis of the tracer is not dependent on the presence of damaged fibers.

By comparing the injection sites of each age group, a position interval can be deduced for the leading descending fibers of each nucleus; this interval is

Table 4. Position intervals of leading descending fibers during development for the various (sub)nuclei. Position intervals were constructed as indicated in Fig. 3

TLD	Lumbosacral	Thoracic	Cervical	MO
E17		▬▬▬▬	▬	
E18		▬		
E19	▬▬▬▬			
E20	▬▬▬▬			
E21	▬▬▬▬			

LC	Lumbosacral	Thoracic	Cervical	MO
E17			▬▬▬	
E18		▬		
E19	▬▬	▬▬		
E20	▬▬▬			
E21	▬▬▬			
P2	▬▬▬			

cNR	Lumbosacral	Thoracic	Cervical	MO
E17				▬▬▬
E18		▬▬▬		
E19		▬▬		
E20		▬		
E21	▬▬▬			
P2	▬▬▬▬			

dNR	Lumbosacral	Thoracic	Cervical	MO
E17				▬▬
E18			▬▬	
E19			▬	
E20			▬▬	
E21		▬▬▬▬	▬	
P2		▬▬▬▬	▬▬	
P3		▬		
P4		▬		

TLD, nucleus tegmentalis laterodorsalis; MO, medulla oblongata; LC, locus coeruleus; cNR, caudal pole of magnocellular nucleus ruber; dNR, dorsal pole of magnocellular nucleus ruber; E17–21, embryonal days 17–21; P2–4, postnatal days 2–4; dmNR, dorsomedial subgroup of NR

located between the rostral border of the caudalmost injection which *did* result in retrograde labeling in the pertinent nucleus and the rostral border of the rostralmost injection which did *not* result in retrograde labeling of this nucleus.

In the fetal spinal cord, (WGA–)HRP injection site borders are diffuse. Because exact determination of the injection site was impossible, the injection site sizes were overestimated, since all sections in which evenly spread, fine-grained TMB deposit was present were included. Thus, the injection site size mainly represents the extent of the area of diffusion of the injected tracer, and not necessarily the extent of the area from which tracer is effectively taken up and transported. The latter area is probably smaller than the former one. The position intervals of the descending fiber fronts (Fig. 3; Table 4) were derived from the positions of the rostral borders of these overestimated injection sites. Since the rostral borders of the effective injection sites are probably located more caudally along the spinal cord, the actual descent (e.g., of the rubrospinal fibers) is probably more advanced than described here and depicted in Table 4.

Axons of serotoninergic neurons in the medullary raphe nuclei were reported to be present at all levels of the spinal cord at E17 and in the dorsal horn of all levels at E19 (Rajaofetra et al. 1989); vestibular axons were reported to be present in the lower spinal cord at E17 (Auclair et al. 1991). The presence of retrogradely labeled neurons in the vestibular and medullary raphe nuclei served as a control for false negative results caused by insufficient transport of the tracer due to imponderabilia.

Even after these precautions, insufficient transport time might still be a source of false negative results, especially so in the case of nuclei located rostrally to the (retrogradely labeled) medullary raphe nuclei. For this reason, series which did not contain retrogradely labeled neurons in the pertinent nucleus were only included if labeled neurons were present at more rostral levels of the central nervous system.

Fibers from the lateral vestibular nucleus are among the earliest descending fibers of the spinal cord, closely followed by fibers from the nucleus raphe magnus and the mesencephalic trigeminal nucleus (Lakke, unpublished observation). Fibers from the medial and descending vestibular nuclei and from the nucleus raphe obscurus follow a few days later. Of the "control nuclei," the nucleus raphe pallidus is the last to initiate descent.

5.4.2 Nucleus Tegmentalis Laterodorsalis; Dorsolateral Funiculus

Descending fibers from the TLD are present in the thoracic spinal cord as early as E17 (Table 4). They enter the lumbar spinal cord during the next day and reach lower lumbar spinal cord levels at E19. TLD fibers have invaded the white matter of the cervical spinal cord by E18 and the white matter of the lumbar spinal cord by E21. Axons from neurons in the ventral TLD appear to be slightly more advanced along the spinal cord than their dorsally originating counterparts. After the arrival of the first fibers at lumbar levels, the establishment of an "adult" distribution pattern of the retrogradely labeled cells in the TLD (actually Barrington's subnucleus) takes 2–3 days.

The presence of only a weak spinobulbospinal reflex at P2 (Kruse and de Groat 1990) might indicate the presence of a substantial delay between the arrival in the target area and the establishment of the first functional synapses. Since the same delay might also be caused by the immaturity of other components of the reflex chain, no firm conclusion can be drawn.

The first fibers appear to originate in the ventral TLD, indicating a ventrodorsal axonogenetic gradient. No specific cytogenetic data are available on the TLD (Altman and Bayer 1980d), preventing the correlation of TLD cytogenesis and axonogenesis.

Descending fibers from the TLD are preceded by vestibulospinal and raphe spinal fibers, but precede those from the LC. Since the earliest fibers from the TLD reach each consecutive spinal cord level both before and after the earliest fibers from nuclei located at the same level of the brainstem (nucleus raphe magnus; LC), the distance between SDPsn (supraspinal descending projection source nucleus) and the spinal cord becomes an unlikely prime determinator of descent chronology.

5.4.3 Locus Coeruleus; Ventral funiculus

At E17, coeruleospinal fibers have reached the upper half of the cervical spinal cord, and at E18 the lower half of the thoracic spinal cord (Table 4). Definite proof of the descent of coeruleospinal fibers down to the upper lumbar spinal cord was obtained at E20, though their presence at upper lumbar levels at E19 cannot be ruled out. Final descent through the lumbar spinal cord is completed in the early postnatal period. A few cells in the rostral LC were retrogradely labeled in the cervical spinal cord at E18, but could not be demonstrated afterwards.

As far as the situation in the P2–P4 time period is concerned, our results conform to those of Leong et al. (1984). These authors also describe the presence of retrogradely labeled neurons in the P2 LC after L3–4 spinal cord injections.

Our results do not support the evidence obtained by Chen and Stanfield (1987) on the presence of a transient projection from the LC to the upper cervical spinal cord, present at P2 and absent at P28 and later; though we did find an indication for the possible presence of a transient projection from the rostral LC present at E18 only, during the E18-P4 period we did not find a high density of retrogradely labeled cells throughout the LC after cervical injections, as Chen and Stanfield did (1987). Similarly, Leong et al. (1984) reported an adult retrograde labeling pattern (i.e., ventrally and caudally in LC) in the P2 rat after C4–5 tracer injections.

The difference may stem from the different tracing techniques used. Leong et al. (1984), like us, used HRP as a tracer, while Chen and Stanfield (1987) employed the fluorescent tracer fast blue (FB). It is conceivable that the initial (distal) segment of translocating axons does not actively endocytoses HRP, but does passively absorp FB, which is fat soluble. Martin and coworkers also reported earlier detection of some connections with the retrograde fluorescent tracer nuclear yellow than with HRP (Martin et al. 1986). On the other hand, it has been shown that active uptake mechanisms are already present in growth cones (Gordon-Weeks et al. 1984; Lockerbie and Gordon-Weeks 1985). Moreover, both of these fluorescent tracers are prone to leakage out of the retrogradely labeled neurons and to subsequent uptake into neighboring neurons (Keizer et al. 1983), which decreases the reliability of the data. Finally, it is equally conceivable that FB (being fat soluble) is absorbed into and diffuses through the degenerating axon shafts of the transient collaterals, even after the cessation of normal retrograde transport, much like the fluorescent tracer DiI (Godement et al. 1987).

Apart from a tracer-technical cause, the difference may also derive from rat species-specific differences which appear to exist in the coeruleospinal projection (Fritschy and Grzanna 1990a,b; Clark and Proudfit 1991). Postmortem tracing with the fluorescent tracer DiI might resolve whether the dorsal LC has a transient spinal projection at birth or not.

Descending coeruleospinal fibers are preceded by the descending fibers from the vestibular and raphe nuclei and from the TLD. Similar observations were made in chick (Okado and Oppenheim 1985). In opossum, however, coeruleospinal fibers were among the earliest fibers present in the thoracic spinal cord, preceding the raphe spinal fibers (Cabana and Martin 1984).

In the rat, the entry of fibers from the LC into the cervical spinal cord occurs about 5 days after the generation of most LC neurons at E12 (Altman and Bayer 1980d). Most of the raphe nuclei neurons are generated at E13, after the LC neurons (Altman and Bayer 1980a,b), but the entry of raphe spinal fibers into the spinal cord occurs at E14 (Rajaofetra et al. 1989), well before the entry of coeruleospinal fibers. These data contradict the notion of SDP source–neuron generation time acting as a prime determinator of descent chronology.

5.4.4 Nucleus Ruber; Dorsolateral Funiculus

At E17, fibers from all subdivisions of the NR have begun their descent through the pons towards the MO and spinal cord (Table 4). Fibers from the cNR and vlNR have reached the caudal MO.

At E18, fibers from the caudal dmNR and the pcNR have reached the caudal MO, while fibers from the vlNR have reached the lower cervical enlargement. Fibers from the cNR have descended to the level of the mid-thoracic spinal cord. In the lower cervical spinal cord, descending rubrospinal fibers are restricted to the white matter.

At E19, fibers from the caudal dmNR have reached the cervical enlargement, and fibers from the cNR and the vlNR have descended to the levels of the lower thoracic spinal cord. Projections from the pcNR have barely reached the cervical spinal cord. In the upper cervical spinal cord, the descending rubrospinal fibers are restricted to the white matter.

At E20, fibers from the dmNR have not reached the lower cervical spinal cord level. Fibers from the vlNR have reached the mid-thoracic level. The most advanced fibers, originating in the cNR, have descended to the levels of the lower thoracic spinal cord.

At E21, fibers from all levels of the dmNR have reached the cervical enlargement. Fibers from the cNR and vlNR have reached the upper lumbar enlargement. Fibers from the pcNR have reached the cervical spinal cord.

Fibers from the cNR and vlNR complete their descent through the lumbosacral spinal cord during the first 4 postnatal days. At P2, fibers from the pcNR have reached the level of the mid-lumbar enlargement. Fibers from the dmNR are present in the lower thoracic spinal cord at P3 or earlier.

The first descending fibers originate in the cNR. These fibers are closely followed by fibers from the vlNR. As soon as these fibers have passed through the cervical enlargement (E19), fibers from the dmNR reach the cervical enlargement. In general, early-descending fibers originate from neurons located caudally and ventrolaterally, and later-descending fibers from neurons located at progressively more rostral and dorsomedial levels of the mcNR, thus leading directly to the adult situation in which the cNR and vlNR project to the lumbar spinal cord and the dmNR to the cervical spinal cord (Brown 1974; Flumerfelt and Gwyn 1974; Reid et al. 1975; Murray and Gurule 1979; Huisman et al. 1981; Shieh et al. 1983; Daniel et al. 1987), These results suggest that the earlier a fiber descends, the further it will reach down the spinal cord, though it should be noted that we do not know whether all neurons within any small area of the NR generate axons at approximately the same time or over a longer time period. Both in the opossum (Xu and Martin 1989) and in the rat (Bregman and Bernstein-Goral 1991), a protracted period of lesion-induced, reactive axonogenesis was described, though regenerating axons, rather than late-growing, axons seem to be involved in the rat.

The descent of the rubrospinal fibers is completed during the first 3 postnatal days. This is in accordance with the results of Shieh et al. (1983) and Leong et al. (1984), who stated that rubrospinal fibers reach lumbosacral levels between P3 and P6.

Injections confined to the gray matter of the spinal cord do not result in retrogradely labeled neurons in any part of the NR prenatally, even though in several experiments (E18 C4162.3, E19 C4011.3) the leading descending fibers from the NR have clearly passed the level of these injections. Apparently, the descending fibers are initially confined to the white matter of the spinal cord and only enter the gray matter of the spinal cord some time after they have reached the appropriate levels of the spinal cord. Such behaviour has been observed in other developing fiber systems as well (Donatelle 1977; Schreyer and Jones 1982; Smith 1983; Cabana and Martin 1986; Martin et al. 1986; Bregman 1987; Rajaofetra et al. 1989). During the prenatal period, injections into the cervical enlargement result (among other things) in retrograde labeling of neurons throughout the cNR. Postnatally, similar injections leave a dorsomedial crescent of the cNR free of retrogradely labeled neurons. Neurons located in this dorsomedial crescent project to the lateral column of the contralateral nucleus facialis (Flumerfelt and Gwyn 1974; Hinrichsen and Watson 1983). It is probable that fibers from these neurons initially bypass the nucleus facialis and descend for some distance into the cervical enlargement of the spinal cord, as do the developing corticopontine projection fibers (Schreyer and Jones 1982). Postnatally, the fibers finally enter the nucleus facialis, and the descending collateral degenerates. This sequence of events would explain the prenatal presence and postnatal absence of retrograde labeling within the dorsomedial crescent of the cNR. Within the spinal cord, a similar sequence of events might take place.

Thus, it seems that the organization of the rubrospinal projection derives from the axonogenetic gradient within the NR. The axonogenetic gradient itself might derive directly from the similarly oriented (caudorostral) neuro-genetic gradient of the NR (Altman and Bayer 1981). Early axons from the caudal and lateral NR are always most advanced along the descent pathway, but reach their target level last. We hypothesize that NR neurons start generating an axon during a fixed period of time after their developmental determination as rubral neurons. These axons descend towards and along the spinal cord along a relatively a-specific "highway" (the central tegmental tract and the lateral funiculi). Final target acquisition of the growth cone (growth perpendicular to the axis of descent and directed into the gray matter) is guided by the concentration gradient of a chemotropic factor which is secreted by the target cells (Heffner et al. 1990; Joosten et al. 1991). Cervical target cells become chemotropic factor secretors before lumbar cells do so. The activity of such a chemotropic factor and its secretion dynamics are currently under investigation.

5.4.5 Cerebral Cortex; Dorsal Funiculus

Corticospinal fibers enter the spinal cord between E21 and P1 and reach the cervical spinal cord at P10 (Gribnau et al. 1986; Schreyer and Jones 1982). The earliest-descending corticospinal fibers originate from lamina V neurons located in the dorsal parietal cortex; these neurons will eventually project into the lumbar spinal cord (Schreyer and Jones 1982). Transient spinal projections are reported from the medial prefrontal cortex (Joosten and van Eden 1989) and posterior (occipital) cortex (Adams et al. 1983; Joosten et al. 1987). Ingrowth of corticospinal fibers into the spinal gray matter occurs an average of 2 days after arrival of the fiber front (Gribnau et al. 1986; Joosten et al. 1987), i.e., cervical at P3, lumbar at P5, and thoracal at P7 (Gribnau et al. 1986). In the cervical spinal cord, a diffusible, chemotropic substance is operative in vitro in cortical growth cone target acquisition (Joosten et al. 1991). Cortical neurons are generated along a coarse lateroventral-to-dorsomedial gradient (Smart 1984; Miller 1988), indicating that forelimb area neurons are probably generated before hindlimb area neurons. The simple (and hypothetical) solution applicable to the development of the rubrospinal tract of first neuron–first fiber and first fiber–last ingrowth is not valid in the case of the corticospinal projection, first since the first fibers seems to stem from the (relatively) last neurons, and second since spinal ingrowth does not seem to follow a spatially continuous gradient. Data on the generation of cortical neurons are, however, relatively imprecise and difficult to coordinate with the orientation of the rattunculus on the somatory–sensory cortex. We are presently studying whether secretion of a diffusible chemotropic substance attracting corticospinal fibers occurs at a different moment in cervical, thoracic, and lumbar spinal cord.

5.4.6 Conclusion

Using our data and data from previous studies, we can now search for patterns in descent chronology (see Table 5). The earliest SDP, which descends through the anterior funiculus of the spinal cord, originates in the lateral vestibular nucleus. These fibers are closely followed by fibers from the adrenergic C1 cells in the nucleus raphe magnus. After a 2-day delay (during which other fiber systems probably descend), fibers from the TLD, LC, NR, and paraventricular hypothalamic nucleus (PVH) start their descent through their respective funiculi at 1-day intervals. Last of all, the corticospinal projection starts descending through the posterior funiculus of the spinal cord.

Fibers from SDP sources located in the proximity of the rhombic lip generally reach the spinomedullary junction before those from SDP sources located further rostrally or caudally. In the latter five SDP, it is clear that the later the SDP arrives in the spinal cord, the longer its descent will last. This is not surprising, since the length of the spinal cord increases (threefold) during the relevant period of development (E16-P10).

In the lateral funiculus, an interesting pattern emerges. The earliest-descending fiber bundle of the lateral funiculus (containing adrenergic and serotonergic fibers) is located in its dorsoventral middle. Subsequently descending fiber systems seem to fasciculate (per system) ventrally and dorsally upon the earlier bundles. In the posterolateral funiculus, the successive fiber systems seem to be crowded by the looming dorsal horn and are forced to curl inward. Similar fasciculations probably occur in the anterolateral and anterior funiculi, involving other descending and ascending systems. These data suggest

Table 5. Neuronal birth dates and descent intervals for various suprasegmental descending projections in the rat

Source	Spinal cord location	Birth	Descent
Adrenergic C1 group	Lateral funiculus	E12–14[a]	E14–17[f]
Vestibular nuclei	Anterior funiculus	E11–15[b]	E14–19[g]
TLD	Posterolateral funiculus	E13–16(?)[a]	E16–19
LC	Anterior/anterolateral funiculi	E12–13[a]	E17–21
NR	Posterolateral funiculus	E13–14[c]	E18–P4[h]
Paraventricular nucleus	Posterolateral funiculus	E13–16[d]	E19–P?[i]
Somatomotor cortex	Posterior funiculus	E15–17[e]	P0–P10[j]

TLD, nucleus tegmentalis laterodorsalis; LC, locus coeruleus; NR, nucleus ruber; E11–21, embryonal days 11–21; P0–10, postnatal days 0–10.
[a] Altman and Bayer 1980d
[b] Altman and Bayer 1980c
[c] Altman and Bayer 1981
[d] Altman and Bayer 1978
[e] Miller 1988
[f] Bernstein-Goral and Bohn 1988
[g] Auclair et al. 1991
[h] Lakke and Marani 1991
[i] Lakke and Hinderink 1989
[j] Gribnau et al. 1986

that time of arrival in the spinal cord determines the position of a SDP system bundle in its respective funiculus. Such an target-independent locator system for the spinal cord funiculi is consistent with the result of developmental lesion studies (Bregman et al. 1989; Schreyer and Jones 1983), which describe aberrant location of SDP system fiber bundles after lesion-induced delay. SDP source neuron birth date obviously limits time of arrival at the spinomedullary junction, but the distance from SDP source to the spinomedullary junction seems to be the prime determinator of the time of arrival at the junction and, thus, of its location in the spinal cord.

6 New Insights into the Development of the Rat Spinal Cord

In the previous chapter, some of the results obtained with the described immuno- and enzyme histochemical techniques shed a new light on our knowledge about the developing rat spinal cord. Several of them will be discussed more extensively in the present chapter. With regard to the developing matrix and mantle layer, the roles of SSEA-1 and AChE will especially be emphasized.

The last part of this chapter will deal with the fusion of the tracing results and the immunocytochemical data concerning the marginal layer. This integration of the different results appeared to be a plausible approach. Beforehand, the results obtained with antibodies against the cytoskeleton markers neurofilaments and MAP seemed most relevant for this approach towards the development of the long supraspinal fibre tracts.

6.1 Matrix Layer

The evident presence of SSEA-1 in the developing matrix layer suggests a close relationship with the spinal cord neuroblasts in general and with the future dorsal horn neurons in particular. Between E9 and E15, cells in the subventricular region of the matrix layer express SSEA-1. This finding is in agreement with previous reports on the presence of SSEA-1 in the developing brain (Marani and Tetteroo 1983; Marani 1986; Yamamoto et al. 1985; Oudega et al. 1992). In the subventricular region, neuroblasts divide before they migrate towards the periphery of the matrix layer. The present results, therefore, indicate an involvement of SSEA-1 in the multiplication of the spinal cord neuroblasts in the dorsal half (alar plate) as well as in the ventral half (basal plate) of the matrix layer. A role in the process of cell division is underlined by the fact that later on, after the pool of spinal cord neuroblasts has been established, the subventricular layer lacks SSEA-1.

Between E13 and E17, the dorsal part of the matrix layer was found to contain SSEA-1. The ventral basal plate, however, appeared to be devoid of this particular molecule. It seems that after the period of cell division, SSEA-1 is especially involved in the further development of those neuroblasts that later settle in the dorsal mantle layer. The exclusiveness of SSEA-1 for this population of neuroblasts indicates a special relationship with these future interneurons. The interneurons function in the conduction of sensory information from the periphery towards the brain. Earlier, a relationship between SSEA-1 and the "sensory" interneurons was suggested in the rat (Dodd and Jessel 1985; Jessel and Dodd 1985) and in the quail (Sieber-Blum 1989).

The possiblity that the distribution of SSEA-1 in the dorsal matrix layer merely reflects its presence on glial cell extensions must be considered. The expression of SSEA-1 on glial cells has been described earlier (Niedieck and Löhler 1987; Mai and Reifenberger 1988; Aloisi et al. 1992). The present study also demonstrates the presence of some radially arranged, vimentin-positive glial fibers in the alar plate between E14 and E17. The radial (vimentin) arrangement does not resemble the overall distribution pattern of SSEA-1 as found in the developing alar plate, which could suggest that the glial processes lack SSEA-1. However, part of the astrocyte population in human spinal cord (Aloisi et al. 1992) and in the adult mouse brain (Bartsch and Mai 1991) express SSEA-1. The latter study demonstrated an abundant presence of the antigen in astrocytes in the barrier structures of the brain (Bartsch and Mai 1991). In the present study, SSEA-1 was demonstrated in the ventral and dorsal midline structures of the rat spinal cord, which could reflect their presence in midline astrocytes. Obviously, all these results should be taken in consideration, and additional experiments are necessary to study the possible colocalization of SSEA-1 and astrocyte markers. One approach could be double staining for SSEA-1 and vimentin.

Another possible function for SSEA-1 could be as an attractant for in-growing fibers. However, around the time of SSEA-1 presence in the dorsal matrix layer (E14–E17), no fiber systems are known to penetrate into the dorsal matrix layer. It seems, therefore, that SSEA-1 is not associated with the ingrowth of fiber tracts into the spinal matrix layer.

The precise function of SSEA-1 in cell division or migration is still obscure. Previously, it was demonstrated that SSEA-1 can be a cell membrane component as well as a cytoplasmatic structure (Mai and Reifenberger 1988). Most likely, the nature of the carrier macromolecule determines the cellular localization of the antigen. Based on this knowledge and the present expression pattern, different functions seem possible for the antigen. Most likely, SSEA-1 is involved in specific migration movements of the neuroblasts towards the periphery of the matrix layer. Another possibility is that SSEA-1 is associated with developmental cellular processes. SSEA-1 has been associated with cell proliferation and differentiation (Düllberg and Mai 1989) as well as with cell–cell contact-mediating processes. Interaction between cells during development requires molecular mechanisms to connect the extracellular environment to cellular sites. Through cell–cell contacts, SSEA-1 and other stage-specific antigens are likely to regulate biological events (Solter and Knowles 1978; Rastan et al. 1985; Niedieck and Löhler 1987; Eggens et al. 1989). This regulatory action is usually triggered by the formation of a complex of the stage-specific antigen and a complementary molecule on the surface of a neighboring cell (the extracellular environment). Directly or indirectly, these complexes may have an effect on intracellular events. In general, cytoplasmatic factors are considered to play regulatory roles in specific gene activities following their entry into the nucleus (Jacobson 1979). Obviously, the environment of a given cell plays an important role in the final determination of its phenotype (see McConnel 1988).

Interestingly, recent studies have showed that developmentally regulated carbohydrates are capable of recognizing their own configuration (Eggens et al.

1989). Consequently, neighboring cells could influence each other directly by recognizing their surface-bound carbohydrate. With regard to the proliferating neuroblasts, it seems possible that SSEA-1 is involved in the initiation/regulation of the division of these subventricularly located cells. In general, matrix cells need living cells to provide the proper environmental signals that regulate their proliferation (Temple 1989). SSEA-1 could function as part of the external signaling that triggers the cells' internal machinery for division, either by binding molecules from the environment and channeling them to the cell cytoplasm or by contacting neighboring cells via surface-bound (complementary) molecules. The above-mentioned knowledge that many carbohydrates are capable of recognizing themselves could be the molecular basis for a direct recognition/influence among the subventricular layer cells, which could result in an intrinsically organized simultaneous cell division.

Later on in development, the maturing future interneurons express SSEA-1 within the alar plate. The neuroblasts are in their mitotic cycles, migrating between the ventricular zone and the midst of the matrix layer. In this stage of development, a role for SSEA-1 could be to ensure the synchronized development of future dorsal horn cells. The presumed ability of SSEA-1 to recognize its own configuration opens the possibility that future dorsal horn cells directly influence each other to achieve simultaneous development. Synchronization could lead to a simultaneous migration of the neuroblasts towards the periphery and coordinated formation of the layers of the dorsal mantle layer. For instance, an organized migration pattern like this could result in a rapid settlement of the neurons in the different dorsal horn laminae (see also Sect. 6.2). Another possible function for SSEA-1 in the migrating future interneurons could be in mediating cell–cell contacts along the migratory pathways, which could determine the morphology and future phenotype of these cells. An involvement of the environment in the final determination of the phenotype of a given cell was recognized earlier (McConnel 1988).

The expression patterns of the enzyme AChE in the developing matrix layer suggest a close relationship with the future spinal cord motor neurons. Between E11–E17, AChE is expressed in the ventral part of the matrix layer of the rat spinal cord, which indicates an exclusive presence of the enzyme in proliferating motor neurons (see also Oudega and Marani 1990). So far, the enzyme has rarely been associated with proliferating neurons of the central nervous system. A small group of "primitive" ChAT-positive cells was recently demonstrated around the ventral half of the mantle layer (Phelps et al. 1991). It appeared that these cells, formerly ventral matrix layer neuroblasts, started to express ChAT as soon as they had left the basal plate. This finding suggests an early presence of the neurotransmitter ACh, although a direct relation between ChAT and ACh was not studied. In the same report, however, photographs of the spinal cord of E12 rat embryos clearly revealed the presence of ChAT in a population of cells within the ventral matrix layer (Phelps et al. 1990). This result, which the authors did not focus on, supports the present finding of a group of cells with a cholinergic character in the basal plate of the rat spinal cord. Therefore, based on the present data as well as these earlier results (Phelps et al. 1990), the future cholinergic neurons of the rat spinal cord seem to be committed to their phenotype even at an early (premigratory) stage of their development. Obviously, such a proposal only holds when AChE does indeed perform a cholinergic function during these early embryonal days (see below).

Contrary to the present results, AChE is not expressed in the chicken in (premigratory) neurons before they reach their final location in the mantle layer (Layer and Sporns 1987). Other reports have mentioned the appearance of BuChE (a pseudo-AChE) in proliferating neurons of the chick nervous system (Layer 1983; Layer et al. 1988). The main difference between these earlier studies and the present one is that in the latter, nonfixed cryostat sections were used to demonstrate the presence of AChE. This could account for the detection of the enzyme in the ventral part of the matrix layer (at all or at an earlier age than previously reported), it has been established that aldehyde fixatives, which are commonly used in anatomical studies on the expression of AChE, are known to reduce the AChE content in the rat central nervous system (McGeer and McGeer 1989). Low amounts of the enzyme could therefore easily be depleted below detection level. It has already been demonstrated that in certain brain areas, AChE can be released from dendrites independent of cholinergic action (Taylor et al. 1988; Greenfield 1991). Another study demonstrated that an increase in neuronal activity results in an increase in the aggregation of AChE subunits (Hüther et al. 1978). Obviously, several studies point to a noncholinergic function of AChE. The early presence of AChE in the ventral matrix layer, as demonstrated in the present study, could also indicate such a noncholinergic role in the development of the premigratory basal plate neuroblasts, i.e., the future motor neurons.

It should be emphasized that the foregoing functional speculations leave unresolved the problem of whether AChE serves as an enzyme in the breakdown of the neurotransmitter ACh or as a regulatory compound by itself. During recent years, the presence of AChE in non-neural cells and in neural cells before actually synthesizing the transmitter ACh has provoked the proposal of several morphogenetic roles for the enzyme in the central nervous system (Kostovic and Rakic 1984; Layer and Sporns 1987). AChE is expected to serve a more metabolic function in the developing brain. Its presence in the ventral matrix layer is yet another finding that suggests a noncholinergic role for the enzyme. If AChE is believed to play a role in the control of the proliferation of the spinal cord motor neurons, it follows from its localization that AChE is active in the final stage of the multiplication process. The present results clearly demonstrate that, contrary to SSEA-1, AChE is not expressed by the cells in the (ventral) subventricular layer, which indicates that AChE is not involved in the initial cell division processes. The exact role of AChE in the later stages of cell proliferation has yet to be elucidated. One possibility is that the presence of AChE in the cell membrane of the motor neuroblasts stimulates the cells towards proper development by channeling the necessary molecular signals from the environment into the cytoplasm. During development, AChE could also have an influence on the cell membrane by a direct action on their channels. Such different functions seem warranted, as recent results demonstrated the capacity of the enzyme to facilitate membrane transport in general and that of sodium in particular (Greenfield 1991; Webb and Greenfield 1992).

A closer anatomical look at the AChE distribution pattern shows that around E16, the enzyme-positive basal plate is sharply delineated from the enzyme-negative alar plate of the spinal cord matrix layer, the latter being SSEA-1 positive as discussed above. The border between the basal and alar plate was found at the level of the sulcus limitans. This indicates the presence and involvement of AChE in the development of all the cells of the spinal cord ventral matrix layer. The existence of a third generation zone within the spinal cord matrix layer, intermediate between the basal and alar plate and producing the future relay neurons, has been proposed (Altman and Bayer 1984). So far, however, conclusive evidence regarding such an additional generation zone is lacking. The present histochemical results do not support the proposal made by Altman and Bayer (1984); they demonstrate a clear twofold subdivision of the matrix layer into an AChE-positive basal plate and a SSEA-1-positive alar plate. However, it is possible that the future spinal cord relay neurons of the intermediate area also require the enzyme for their proper development and that, consequently, this additional generation zone does not stand out as a separate area after an AChE staining.

It must also be noted that for a short period of time (E12–E13), SSEA-1 transgressed the level of the sulcus limitans. This could suggest the presence of SSEA-1 in a third (intermediate) region besides the dorsal (positive) alar plate and the ventral (negative) basal plate. The short time frame of 2 days corresponds with the period of peak production of the relay cells, which are thought to be produced in the intermediate region between E11 and E14 (see Altman and Bayer 1984). Its absence before and after that period does not support a function of SSEA-1 in the production of those cells. Future research should clarify whether SSEA-1 has a particular short-term role in the interneurons that are generated in that region.

The above-mentioned and more extensively discussed results of the two studied functional morphoregulators SSEA-1 and AChE clearly demonstrate that the spinal cord matrix layer contains histochemically distinguishable cell populations. Apparently, the matrix layer is not a homogeneous population of cells, as has been thought for a long time. Other recent evidence for its heterogeneity was provided by the demonstration of several distinct cell groups located along the lumen of the matrix layer (Silos-Santiago and Snider 1990). The presence of a number of histochemically distinguishable cell populations seems acceptable in view of the developmental pathways followed by the different types of matrix layer cells. These separate cell populations obviously need different molecular signals that indirectly or directly trigger certain events during particular periods. Ultimately, such an organized maturation of these different cell groups will lead to the formation of functionally integrated cell populations in the spinal cord mantle layer.

6.2 Mantle Layer

During the development of the mantle layer of the rat spinal cord, some particularly surprising events are found in the dorsal part. From E18 on, SSEA-1 is present in three separate areas in the dorsal horn. So far, SSEA-1 has not been demonstrated in the mantle layer, which has led to the suggestion that migratory or postmigratory neurons lack this particular molecule (Yamamoto et al. 1985).

Although difficult to determine, the three separate SSEA-1 positive areas in the dorsal mantle layer can be identified in adjacent cresyl violet-stained sections as Rexed's laminae II, IV, and VI/VII (including the lateral horn). During further development, the molecule disappears from laminae IV and VI/VII. Finally, in the mature spinal cord, only lamina II remains positive for SSEA-1. It seems, therefore, that its role in the development of the inter-neurons of lamina II is more widespread than its role in the development of the relay neurons of the other two regions.

Based on its transient presence in two of the three areas, several functions for SSEA-1 can be suggested. Firstly, SSEA-1 could be involved in developmental events such as the regulation or modification of the migration and settlement of dorsal horn neurons. Such an involvement of SSEA-1 seems acceptable, other stage-specific antigens are also implicated in guidance functions (Solter and Knowles 1978; Rastan et al. 1985; Niedieck and Löhler 1987; Eggens et al. 1989; Plank and Mai 1992). It has already been mentioned that, based on the present results, an involvement of SSEA-1 in migratory movements of the future dorsal horn (inter)neurons appears to be a possibility. Recently, it was found that other carbohydrates active in developmental events are capable of recognizing their own configuration (see Eggens et al. 1989). Consequently, SSEA-1-positive cells could directly influence each other, which could lead to an organized migration and settlement of these cells in the mantle layer. Because of the transient expression of SSEA-1, one could state that the settlement of the neurons in lamina VI/VII is completed during the first 2 days after birth and slightly ahead of that in lamina IV. This would be in total agreement with the general ventral-to-dorsal gradient in the developing spinal cord.

The second possible function of SSEA-1 could be a role in the synchronization of the development of the neuronal population of laminae IV and VI/VII. The presumed ability of SSEA-1 to recognize its own configuration, like other developmentally regulated carbohydrates (Eggens et al. 1989), underlines such a synchronization function, since the positive neurons are able to influence each other directly. The advantage of simultaneous development is that most of the cells within a certain population are in the proper developmental stage at the time of arrival of the afferents (from the periphery and from the higher brain center). A recent study demonstrated that the ingrowth of dorsal root fibers in the dorsal mantle layer largely takes place during the last few embryonal days (Ruit et al. 1992). Besides receiving fibers from the sensory cortex (lamina IV), the motor cortex, and the nucleus ruber (lamina VI/VII; Nyberg-Hansen and Brodal 1963), both laminae are also known to receive input from primary afferents (Brown 1981). The period during which the primary afferents enter the dorsal mantle layer corresponds well with the period in which the receiving cells express SSEA-1. The presence of SSEA-1-complementary structures on the surface of the ingrowing primary afferents would guarantee the recognition of their target cells and, consequently, the proper development of the sensory system. A possible scenario could be that the ingrowing primary afferents express SSEA-1, which recognizes the SSEA-1 structure present on the dorsal horn neurons. Such a scenario seems plausible, as presumably SSEA-1 is capable of recognizing its own configuration.

The persistence of SSEA-1 in lamina II beyond the period of settlement and receiving the afferent input implies that the above-mentioned (transient) roles are only part of the complete picture of SSEA-1 in relation to its function in the interneurons of this specific lamina. The different suggestions concerning a role in the migration and settlement or in a synchronized development of this particular neuronal population are feasible, as the molecule is clearly present during these events, but clearly its expression in the adult rat in this particular lamina must serve other goals. Lamina II of the rat dorsal horn is known to be part of the substantia gelatinosa in which the unmyelinated primary afferents find their appropriate target cells. The exclusiveness of SSEA-1 for the dorsal neuroblasts (see above) and its specific localization in the adult substantia gelatinosa may indicate a close relationship not only between the antigen and the development of the C fibers and A(∂) fibers of the rat sensory system, but also their functioning (Dodd and Jessel 1985; Jessel and Dodd 1985; Oudega et al. 1992). In addition to the establishment of their synaptic contacts with the interneurons of lamina II, the maintenance of these synapses could also be regulated by SSEA-1. Earlier similar conclusions were drawn from an immunocytochemical study in which the mature rat spinal cord was found to contain several examples from the group of lactoseries carbohydrates (Dodd and Jessel 1985).

6.3 Marginal Layer

The description of NF-90 immunoreactivity in the marginal layer of the spinal cord during development provides a general background for the systematic description of the SDP timetable. NF-90 antibodies cannot distinguish descending, ascending, and propriospinal fibers, but label all fibers that contain phosphorylated neurofilament proteins (Oudega et al. 1992). Axons can be labeled anterogradely throughout translocation (Cholley et al. 1989; Crabtree 1990; Pippenger et al. 1990; Schreyer and Jones 1982, 1983, 1988; Thong and Dreher 1987). Even though several studies exist that employ retrograde transport of HRP from the spinal cord (Chen and Stanfield 1987; Leong 1983; Leong et al. 1984; Nordlander et al. 1985; Okado and Oppenheim 1985), we do not know for sure whether the initial segment of the translocating axon is capable of retrograde HRP transport or whether the pioneer axons (if present) are capable of retrograde HRP transport. However, anterograde and retrograde developmental studies of the corticospinal system are in complete agreement on the arrival time of the fiber front at various levels (Donatelle 1977; Leong 1983), indicating that retrograde HRP transport is indeed possible from all· sections of the translocating axon. Thus, the presence of NF-90 immunoreactivity in fibers along the route of the corticospinal fibers at E20, before cortical neurons can be retrogradely labeled from the spinal cord, is evidence of the presence of another fiber system, separate from the corticospinal system, possibly descending fibers from the cuneate and gracile nuclei (Leong et al. 1984).

Though the resolution of NF-90 immunocytochemistry was not high enough to discriminate the various smaller SDP systems, the corticospinal and rubrospinal systems are easily distinguished. If anything, NF-90 immunoreactivity in the corticospinal and rubrospinal tracts seems to be delayed with respect to the arrival time of the descending fibers, as evidenced from the HRP studies. Though axons growing from cortical and rubral explants, cultured in three-dimensional collagen matrices, express NF-90 immunoreactivity both in the axon shaft and the growth cone at least from the moment they emerge from the explant (Lakke, unpublished observation), it can be argued that the cultured developing axons are in fact different from the "real" developing axons, since they emerge from a neuron which has had a former axon severed. NF-90 immunoreactivity in an axonal tract thus reflects the state of its originating neurons and is, at least in the cases of the corticospinal and rubrospinal tracts, only present in the tract at a certain spinal cord level some time after the fiber front has passed through this level. The state of the neuron in turn is most probably determined by its age or maturity, but might also represent other parameters such as the length of axon generated or the presence of afferent synaptic contacts.

The NF-90-immunoreactive fibers present in the fasciculus gracilis at lumbar and thoracic levels at E18 most probably represent ascending fibers from the lower spinal ganglia (Wessels 1991). Since fibers from the lower ganglia have to travel the longest distance through the (ever-elongating) spinal cord, their early departure would be in agreement with the principles observed in the SDP systems. However, the necessary axonogenetic gradient (caudal-to-rostral) is contrary to experimental data (Fitzgerald et al. 1991) and contrary to the neurogenetic gradient (rostral-to-caudal) of the spinal cord ganglia (Altman and Bayer 1984). The fact, however, that the axons from the youngest (caudal) ganglionic neurons are the first to express NF-90 immunoreactivity again indicates that neurofilament phosphorylation is not a simple maturational parameter, but rather a functional parameter, reflecting, for instance, the length of the axon generated by the originating neuron.

Whatever its significance with respect to the state of the originating neuron, the presence of NF-90 immunoreactivity in the spinal cord white matter during the early stages of development does at least provide evidence of the presence of axons. Though timing the arrival of axons at a certain level of the spinal cord is not possible with NF-90 immunoreactivity (due to the expressional delay), the sequence of arrival seems adequately reflected. Once the spinal cord white matter has filled up with NF-90 immunoreactivity, its interpretation becomes more difficult. Postnatally (>P4), all fiber systems (with the exception of the corticospinal tract) have completed descent. Though some addition of fibers occurs for a longer period of time (Bregman and Bernstein-Goral 1991) and though there is a significant degeneration of SDP fibers in the spinal cord at later developmental stages (Gorgels 1990), these events cannot be related to NF-90 immunoreactivity shifts occurring in the superficial anterior and lateral funiculi. In these later stages of development, neurofilament phosphorylation apparently reflects yet another functional parameter, whose nature remains elusive. This implies that the presence of NF-90 immunoreactivity represents different neuronal states at different stages of development. (To complicate matters, not all levels of the spinal cord are synchronized with respect to developmental stage.) During prenatal development, NF-90 immunoreactivity reflects neurofilament phosphorylation necessary for the stabilization of a translocating axon. This stabilization occurs first when a certain length of axon is generated, in the proximal axonal segment. Failing the generation of a sufficiently long axon (if source and target are close together), stabilization occurs after arrival in the target area, in the whole axon. During later development (>P4), some reorganization takes place, involving the dephosphorylation and phosphorylation of neurofilament proteins and possibly the disassembly of NF-L and NF-M polymers in favor of the more stable NF-H polymers (Nixon and Shea 1992).

7 Summary

The reciprocal cooperation between the genome and epigenetic factors determines the final outcome of the development of the rat spinal cord. This study presents several developmentally regulated epigenetic factors involved in the development of the neuronal and glial cell population of the rat spinal cord. By correlating their appearance and distribution patterns with known developmental events, possible roles of the factors are suggested and discussed. Neuronal markers such as neurofilaments, MAP 1A, 2 and 5, SSEA-1, and the enzyme AChE were studied during the development of the spinal cord. The glial cell markers vimentin and GFAP were also examined. In addition, the development of several major supraspinal fiber systems was studied with WGA–HRP as neuronal tracer. This summary presents only the main results for the three different spinal cord layers. Others are mentioned and, together with the former, more extensively discussed in Chaps. 5 and 6.

7.1 Matrix Layer

Between E12 and E17, the developing matrix layer displayed radially orientated MAP 1A- and MAP 5-positive fibers. During this period, the appearance and disappearance of these MAP-positive processes followed a ventral-to-dorsal gradient. In the same time frame and according to a similar gradient, proliferating spinal cord neuroblasts originate in the subventricular region and migrate back and forth along earlier-extended cytoplasmatic processes. Most likely, the temporary presence of MAP 1A and MAP 5, which are both involved in the assembly and stabilization of microtubules, reflects their involvement in the organization of the (dynamic) cytoskeleton of the neuroblast extensions.

Between E9 and E15, SSEA-1 was present in the subventricular region of the spinal cord matrix layer. Its presence indicates an involvement of SSEA-1 in the division of the neuroblasts, which is known to take place in the small region adjacent to the central lumen. Additionally, between E13 and E17, SSEA-1 was expressed in the developing alar plate. In the alar plate, the dorsal part of the matrix layer, the future neurons of the dorsal horn are generated. Its temporary presence suggests a role of SSEA-1 in the development of the dorsal neuroblasts, most likely in their migration towards the periphery of the matrix layer.

Between E11 and E16, the developing basal plate, the ventral part of the matrix layer, contained AChE. After E16, the enzyme rapidly disappeared from the basal plate. This temporary presence indicates an involvement of AChE in the proliferation of the future spinal cord motor neurons which originate from the ventral basal plate.

Between E11 and P4, the glial cell marker vimentin was expressed throughout the matrix layer, first scattered all over and later on in a pattern of radially arranged processes. The appearance and disappearance of these glial fibers followed a ventral-to-dorsal gradient. Within the spinal cord matrix layer, the neuroblasts are known to be generated according to a similar ventral-to-dorsal gradient. The temporary presence of the radially orientated glial fibers could reflect their involvement in the migration of the neuroblasts towards the periphery. During the first 3 postnatal weeks, the major presence of vimentin-positive processes was replaced by a minor presence of long, thin, GFAP-positive glial fibers. These latter fibers formed the ventral and dorsal midline in the adult spinal cord.

7.2 Mantle Layer

Around E12–13, the first phosphorylated neurofilaments were present in the ventral part of the developing mantle layer. During the following days, neuro-filaments appeared according to a ventral-to-dorsal gradient, which coincided nicely with the formation of the different regions in the mantle layer. Apparently, phosphorylated neurofilaments are only expressed in postmitotic neurons just after their arrival in the periphery of the cord and just before the start of the outgrowth of their processes. In general, the mantle layer cells and their fibers only faintly expressed neurofilament subunits. After birth, the neurofilaments gradually vanished from the mantle layer cells towards a faint presence. In the adult spinal cord, only sparse, ventrally located motor neurons exhibited a moderate neurofilament staining intensity. The overall faint presence of neurofilaments is probably due to the specific character of the used antibody, NF-90. This antibody recognizes the phosphorylated forms of the neurofilament subunits, which are mostly located in the axon (see Sect. 7.3). Because the presence of the neurofilaments could not be directly correlated to specific developmental events in the spinal cord, their role seems restricted to being a member of the axonal cytoskeleton.

At E12, MAP 1A and MAP 5 were first found in cells and fibers in the ventral mantle layer. During the next few days, both proteins were expressed according to a ventral-to-dorsal gradient, nicely following the formation sequence of the expanding mantle layer. Between E12 and E16, the distribution patterns indicated that the motor neurons as well as the contralaterally projecting (intermediate gray) neurons require MAP 1A and 5 for the organization of their developing microtubules. After birth, the intensity of MAP staining gradually diminished towards a faint appearance at maturity. The temporary presence of MAP 1A and MAP 5 clearly underlines their particular involvement in the growing phase of the fibers and is in agreement with their function in the process of assembly and stabilization of microtubules. Nevertheless, both MAPs are still faintly present in the adult spinal cord throughout the mantle layer. Such a low level of these "developmental" proteins could reflect the presence of a certain degree of neuronal plasticity. Between E12 and E15, MAP 2 is expressed in the developing mantle layer according to a similar ventral-to-dorsal gradient as the one mentioned above. Its appearance, therefore, reflected the arrival and development of the neuroblasts in the different regions in the mantle layer. After birth, the MAP 2 staining intensity gradually declined towards a faint level in the adult mantle layer. The overall presence of MAP 2 in the developing and mature mantle layer demonstrates that besides its involvement in the organization of microtubules in growing dendrites, MAP 2 also plays a role in the maintenance of the adult dendritic microtubules. Contrary to the faint presence in the adult mantle layer, MAP 2 appeared to be moderately expressed in Rexed's laminae I and II of the dorsal horn. This could reflect the presence of a high number of dendrites in these two layers, which are known to receive incoming primary afferents from the periphery.

From E17 on, three different regions in the dorsal part of the mantle layer contained SSEA-1. During the following days, the antigen gradually vanished from the two ventrally located regions, but it persisted in the most dorsal one. The remaining positive region was identified in adjacent cresyl violet-stained sections as Rexed's lamina II (substantia gelatinosa). At maturity, SSEA-1 was still present in lamina II, which receives incoming primary afferents. The findings underline the close relationship of SSEA-1 with the developing and mature spinal cord sensory system. The distribution pattern of SSEA-1 indicates and involvement in the migration and organization of the neurons of the dorsal horn as well as in the formation and maintenance of part of their connections. After birth, coarse SSEA-1 staining could be detected throughout the mantle layer, which most likely reflects its presence in constituents of the myelin sheaths. The myelination of the fiber systems in the spinal cord is known to take place during the first few postnatal weeks.

Between E12 and E17, the enzyme AChE was present within the ventral mantle layer in developing motor neurons and their axons. The temporary presence of the enzyme suggests an association with developmental events such as cell clustering (motor columns) and fiber pathfinding (motor roots). After birth, a faint presence of AChE was detected throughout the mantle layer. However, during the second postnatal week, the enzyme was especially displayed in Rexed's lamina II (substantia gelatinosa), which could reflect a late arrival of primary afferents. From E15 on, AChE was markedly present in the intermediate gray of the developing mantle layer. The intermediolateral and intermediomedial cell groups and the intercalated region in between them (all three part of the autonomic system) contained AChE-positive cells and fibers. After birth, however, most of the enzyme activity gradually disappeared from the intermediomedial cell group and the intercalated region. This temporary presence suggests a role for AChE in the development of the rat autonomic system. In the adult rat spinal cord, AChE was still detected in the intermediolateral cell clusters (located in the lateral horn), most likely reflecting their cholinergic nature.

Both vimentin and GFAP appear to contribute to the basic structural organization of the spinal cord mantle layer. Between E13 and E18, a radial distribution pattern of vimentin-positive glial fibers was demonstrated in the developing mantle layer. This vimentin distribution pattern suggests an involvement of glial fibers in the migration of neuroblasts from the matrix layer towards their proper location in the mantle layer. During the first 3 postnatal weeks, vimentin gradually vanished and GFAP gradually appeared in the spinal cord mantle layer. Contrary to the overall tangential orientation, the GFAP-positive glial fibers in the dorsal part of the mantle layer coursed in a ventral-to-dorsal direction. This could reflect the involvement of these glial fibers in the guidance of the incoming primary afferents towards the upper layers of the dorsal horn.

7.3 Marginal Layer

From E13 on, neurofilaments were found to be expressed in longitudinally orientated fibers in the ventral funiculus. Positive fibers were also detected in the ventral commissure. This distribution pattern demonstrated the presence of (ascending) contralaterally projecting fibers and/or (descending) supraspinal tracts. One day earlier, however, longitudinal, neurofilament-positive fibers were already present in the lateral funiculus, suggesting that the ipsilaterally projecting fibers develop slightly ahead of the contralaterally projecting axons. A different explanation for this time delay of about 1 day could be the larger distance that the contralaterally projecting fibers have to travel to reach the ventral funiculus. If this is the case, the two different cell populations of the intermediate gray are generated and develop simultaneously. Developmental studies on a variety of other species suggest that descending supraspinal tracts and propriospinal projections are likely to be among the neurofilament-positive fibers in the lateral funiculus during these early days of embryonal development. Between E13 and E16, more intense neurofilament staining was found in the periphery of the lateral funiculus. Later on in development, a concentration of neurofilaments was found in the subsurface region. This distribution pattern could reflect a developmental phenomenon in which the earlier-generated fibers are being pushed medially by the later-generated fibers. In most cases, it was unclear whether the demonstrated concentration of neurofilaments is the result of a higher number of fibers or of the presence of more phosphorylated neurofilaments per (older) fiber.

At E13, the dorsal root entrance zone and the dorsal root bifurcation zone contained neurofilament-positive, longitudinally coursing fibers, demonstrating primary afferents penetrating the spinal cord. Around E16, the initial dorsal funiculus, which has developed on top of the dorsal horn, displayed more neurofilaments in its most medial tip, and at E18 a concentration of neurofilaments was present in the dorsal part of the fasciculus gracilis. The presence of a higher number of fibers in these particular regions could be responsible for this concentration of neurofilaments, but the possibility of the presence of more phosphorylated neurofilament subunits per fiber must also be considered. Around E18, the first (pioneer) fibers were found to be present in the corticospinal tract region (the most ventral part of the dorsal funiculus). This area filled with fibers after birth according to a rostral-to-caudal gradient. The lumbar levels were reached around P12, indicating the earlier-described extended development of the rat corticospinal tract.

After E13, longitudinally coursing MAP 1A- and MAP 5-positive fibers were found to be prcsent in the lateral and ventral funiculus. After birth, both MAP gradually disappeared from these fibers. The appearance of the MAP coincides with the expression of neurofilaments, reflecting the coordinated production of the cytoskeleton constituents. The temporary presence of MAP 1A and MAP 5 indicates their involvement in the development of the ipsilaterally and contralaterally projecting fibers of the rat spinal cord. Additionally, supraspinal and propriospinal axons are likely to be among the MAP-positive, longitudinally orientated fibers.

From E13 on, MAP 1A- and MAP 5-positive, longitudinal fibers were present in the dorsal root entrance zone and the dorsal root bifurcation zone. From E17 on, the fibers in the fasciculus gracilis of the dorsal funiculus were found to contain MAP 5, but appeared to be devoid of MAP 1A. The fibers in the fasciculus cuneatus contained both MAP. The lack of MAP 1A in the fibers of the fasciculus gracilis most likely reflects the fact that these ascending fibers do not need this particular MAP. After birth, both MAP gradually disappeared from the dorsal funiculus, indicating their role in the growing phase of the fibers. During development, the fibers in the corticospinal area were found to contain MAP 5. These descending fibers only faintly expressed MAP 1A. After birth, however, MAP 1A appeared to be abundantly expressed, whereas MAP 5 was only faintly present. This finding indicates that the descending corticospinal fibers are in need of MAP 5 during their growing phase, which is in agreement with its presumed role in the assembly and stabilization of microtubules. In contrast to MAP 5, MAP 1A is not required by the fibers during their development, but this molecule is needed to maintain the adult organization of the microtubules.

During the embryonal period, the spinal cord funiculi lacked SSEA-1, except for a small region in the developing lateral funiculus during E13-E14. This positive region corresponded with the dorsal root entrance zone. Its presence in this particular region underlines the involvement of SSEA-1 with the ingrowing primary afferents. A possible role for SSEA-1 at this moment of development could be the initial guidance and/or attraction of the ingrowing primary afferents towards the dorsal root bifurcation zone. After birth, SSEA-1 was found to be expressed in a coarse pattern throughout the spinal cord funiculi (including the dorsal funiculus and the corticospinal tract area, but excluding the dorsolateral fasciculus). This particular distribution pattern suggests the presence of SSEA-1 in the myelin sheaths, which are formed after birth.

Except for a small group of cells in the dorsal part of the lateral funiculus, the enzyme AChE could not be demonstrated in the spinal cord funiculi. The AChE-positive cells most likely belong to the lateral spinal nucleus, although at the thoracic level (AChE positive) autonomic cells could be intermingled. From E13 on, AChE is present in the dorsal root entrance zone and the dorsal root bifurcation zone. During further development, the enzyme activity gradually disappeared from the dorsal funiculus. This temporary presence indicates an involvement of AChE in thc growing phase of the ascending sensory fibers. One of the more likely roles for AChE is in the pathfinding of the developing axons. Unlike the ascending sensory fibers, the descending corticospinal axons did not express the enzyme. Throughout development and in the adult rat spinal cord, the corticospinal tract region appeared to be devoid of AChE.

From E14 on, the different funiculi (including the corticospinal tract area) of the spinal cord displayed palisades of vimentin-positive fibers. These glial fibers most likely fulfil a role in the guidance of the growing ascending/descending fiber tracts. During the first few weeks after birth, vimentin presence is replaced by GFAP presence. GFAP was found to be organized in similar palisades of (radially orientated) fibers. Most likely, the arrangement of glial fibers in the adult cord primarily contributes to the structural organization of the rat spinal cord funiculi.

The neuronal tracer WGA–HRP was used to study the development of several descending supraspinal fiber systems. By E17, fibers of the coeruleospinal tract have reached the upper cervical levels of the cord. One day later, the lower thoracic levels were innervated, and around E20, the lumbar levels. The results demonstrate that the coeruleospinal fibers need 3 days to reach the lowest levels of the spinal cord. Descending fibers from the TLD reached the thoracic spinal cord at E17. One day later, these fibers entered the higher lumbar levels, and at E19, the lowest lumbar levels. The cervical white matter was invaded by tegmentospinal fibers at E18, and the lumbar white matter, around E21. The findings suggest that the tegmentospinal fibers reach the lower spinal cord levels within 2 days. Additionally, the results indicate a delay of 2 days between arrival and invasion of tegmentospinal fibers of the different spinal cord levels.

Fibers from the cNR have reached the mid-thoracic levels of the spinal cord by E18 and the lower thoracic levels by E19. Fibers from the vLNR were slightly behind, but also reached the lowest thoracic levels by E19. At E19, fibers from the dmNR entered the mid-cervical levels. By E21, the axons originating in the cNR and vLNR have reached the upper lumbar levels, and around P4, the lowest lumbar spinal cord levels. In contrast to the other two fiber systems examined, the results indicate an extended development of the rat rubrospinal axons. During the first few postnatal days, after a delay of approximately 2–3 days, the rubrospinal fibers started the invasion of the gray matter. An axonogenetic gradient in the organization of the rubrospinal projections is proposed.

References

Abney ER, Bartlett PP, Raff MC (1981) Astrocytes, ependymal cells, and oligodendrocytes develop on schedule in dissociated cell cultures of embryonic rat brain. Dev Biol 83:301–310

Adams CE, Mihailoff GA, Woodward DJ (1983) A transient component of the developing corticospinal tract arises in the visual cortex. Neurosci Lett 36:243–248

Aguayo AH, David S, Bray GM (1981) Influences of the glial environment on the elongation of axons after injury: transplantation studies in adult rodents. J Exp Biol 95:231–240

Aloisi F, Giampaolo A, Russo G, Peschle C, Levi G (1992) Developmental appearance, antigenic profile, and proliferation of glial cells of the human embryonic spinal cord: an immunocytochemical study using dissociated cultured cells. Glia 5(3):171–181

Altman J, Bayer SA (1978) Development of the diencephalon in the rat. I. Autoradiographic study of the time of origin and settling patterns of neurons in the hypothalamus. J Comp Neurol 182:945–972

Altman J, Bayer SA (1980a) Development of the brainstem in the rat. I. Thymidine-radiographic study of the time of origin of neurons of the lower medulla. J Comp Neurol 194:1–35

Altman J, Bayer SA (1980b) Development of the brainstem in the rat. II. Thymidine-radiographic study of the time of origin of neurons in the upper medulla, excluding the vestibular and auditory nuclei. J Comp Neurol 194:37–56

Altman J, Bayer SA (1980c) Development of the brainstem in the rat. III. Thymidine-radiographic study of the time of origin of neurons of the vestibular and auditory nuclei. J Comp Neurol 194:877–904

Altman J, Bayer SA (1980d) Development of the brainstem in the rat. IV. Thymidine-radiographic study of the time of origin of neurons in the pontine region. J Comp Neurol 194:905–929

Altman J, Bayer SA (1981) Development of the brainstem of the rat. V. Thymidine-radiographic study of the time of origin of neurons in the midbrain tegmentum. J Comp Neurol 198:677–716

Altman J, Bayer SA (1984) The development of the rat spinal cord. In: Beck et al. (eds) Advances in anatomy, embryology and cell biology, vol 85. Springer, Berlin Heidelberg New York

Anderton BH, Brion J-P, Flament-Durand J, Haugh MC, Kahn J, Miller CCJ, Probst A, Ulrich J (1987) Neurofibrillary tangles and the neuronal cytoskeleton. J Neural Trans 24:191–196

Andrezik JA, Beitz AJ (1985) Reticular formation, central gray and related tegmental nuclei. In: Paxinos G (ed) The rat nervous system, vol 2. Academic, Sidney, pp 1–28

Angaut P, Batini C, Billard JM, Daniel H (1986) The cerebellorubral projection in the rat: retrograde anatomical study. Neurosci Lett 68:63–68

Angulo Y, Gonzáles AW (1940) The differentiation of the motor cell columns in the cervical cord of albino rat fetuses. J Comp Neurol 73:469–488

Arsénio–Nunes ML, Sotelo C (1985) Development of the spinocerebellar system in the postnatal rat. J Comp Neurol 237:291–306

Auclair F, Belanger MC, Marchand R (1991) Ontogenesis of descending pathways from various brainstem nuclei to the spinal cord in the rat. Soc Neurosci Abstr 17:41

Autilio-Gambetti L, Crane R, Gambetti P (1986) Binding of Bodian's silver and monoclonal antibodies to defined regions of human neurofilament subunits: Bodian's silver reacts with a highly charged unique domain of neurofilament. J Neurochem 46:366–370

Barakat-Walter I, Riederer BM (1991) Brain spectrin 240/235 and 240/235E: differential expression during development of chicken dorsal root ganglia in vivo and in vitro. Eur J Neurosci 3:431–440

Barrington FJF (1925) The effect of lesion of the hind- and midbrain on micturition in the cat. J Exp Physiol 15:81–102

Bartsch D, Mai JK (1991) Distribution of the 3-fucosyl-N-acetyl-lactosamine (FAL) epitope in the adult mouse brain. Cell Tissue Res 263(2):353–366

Basbaum AI, Hand PJ (1973) Projections of cervicothoracic dorsal roots to the cuneate nucleus of the rat with observations on cellular "bricks." J Comp Neurol 148:347–360

Baumal R, Kahn HJ, Bailey B, Philips MJ, Hanna W (1980) The value of immunohisto-chemistry in increasing diagnostic precision of undifferentiated tumours by the surgical pathologist. Histochem J 16:1061–1078

Bear MF, Carnes KM, Ebner FF (1985) Postnatal changes in the distribution of acetyl-cholinesterase in kitten striate cortex. J Comp Neurol 237:519–532

Bentley D, Caudy M (1983) Pioneer axons lose directed growth after selective killing of guidepost cells. Nature 304:62–65

Benviniste EN, Whitaker JN, Gibbs DA, Sparacio SM, Butler JL (1989) Human B cell growth factor enhances proliferation and glial fibrillary acidic protein gene expression in rat astrocytes. Int Immunol 1(3):219–228

Bergquist H (1964) The formation of the front part of the neural tube. Experientia 20:92–97

Bernhardt R, Matus A (1984) Light and electron microscopic studies of the distribution of microtubule-associated protein 2 in rat brain: a difference between the dendritic and axonal cytoskeletons. J Comp Neurol 226:203–221

Bernhardt R, Huber G, Matus A (1985) Differences in the developmental patterns of three microtubule-associated proteins in the rat cerebellum. J Neurosci 5:977–991

Bernstein-Goral H, Bohn MC (1988) Ontogeny of adrenergic fibers in rat spinal cord in relationship to adrenal preganglionic neurons. J Neurosci Res 21:333–351

Bhat S, Pfeiffer SE (1986) Stimulation of oligodendrocytes by extracts from astrocyte-enriched cultures. J Neurosci Res 15:19–27

Bignami A, Dahl D (1987) Axonal maturation in development. II. Immunofluorescense study of rat spinal cord and cerebellum with axonal-specific neurofilament antibodies. Int J Neurosci 5:29–37

Bignami A, Raju T, Dahl D (1982) Localization of vimentin, the nonspecific intermediate filament protein, in embryonic glia and in early differentiating neurons. Dev Biol 91:286–295

Binder LI, Frankfurther A, Kim H, Caceres A, Payne MR, Rebhun LI (1984) Heterogeneity of microtubule-associated protein 2 during rat brain development. Proc Natl Acad Sci USA 81:5613–5617

Bloom GS, Schonfeld TA, Vallee RB (1984) Widespread distribution of the major polypeptide component of MAP1 (microtubule-associated protein 1) in the nervous system. J Cell Biol 98:320–330

Blumenkopf B (1988) Neurochemistry of the dorsal horn. Appl Neurophsiol 51:89–103

Bodian D (1936) A new method for staining nerve fibres and nerve endings in mounted paraffin sections. Anat Rec 65:89–97

Bok ST (1928) Das Rückenmark. In: Von Möllendorff W (ed) Handbuch der mikroskopischen Anatomie des Menschen. Springer, Berlin, pp 478–578

Bondar LR (1977) Development of glia. In: Schoffeniels et al. (eds) Dynamic properties of glia cells. Pergamon, Oxford, pp 3–13

Bovolenta P, Liem RKH, Mason CA (1984) Development of cerebral astroglia transitions in form and cytoskeletal content. Dev Biol 102:248–259

Bovolenta P, Liem RKH, Mason CA (1987) Glial filament protein expression in astroglia in the mouse visual pathway. Dev Brain Res 33:113–126

Bray D (1982) The mechanism of growth cone movements. Neurosci Res Prog Bull 20:821–829

Bregman BS (1987) Development of serotonin immunoreactivity in the rat spinal cord and its plasticity after neonatal spinal cord lesions. Dev Brain Res 34:245–263

Bregman BS, Bernstein-Goral HB (1991) Both regenerating and late-developing pathways contribute to transplant-induced anatomical plasticity after spinal cord lesions at birth. Exp Neurol 112:49–63

Bregman BS, Kunkel-Bagden E, McAtee M, O'Neill A (1989) Extension of the critical period for developmental plasticity of the corticospinal pathway. J Comp Neurol 282:355–370

Brodal A (1981) The dorsal roots and afferent fibres in the ventral roots. In: Brodal A (ed) Neurological anatomy, 3rd edn. Oxford University Press, Oxford, pp 60–69

Brown AG (1981) Organization in the spinal cord. Springer, Berlin Heidelberg New York

Brown BL, Epema A, Marani E (1986) Topography of acetylcholinesterase in the developing rabbit and cat cerebellum. Prog Histochem Cytochem 16(4):117–127

Brown LT (1971) Projections and terminations of the corticospinal tract in rodents. Exp Brain Res 13:432–450

Brown LT (1974a) Corticorubral projections in the rat. J Comp Neurol 154:149–167

Brown LT (1974b) Rubrospinal projections in the rat. J Comp Neurol 154:169–188

Burnside B, Jacobson AG (1968) Analysis of morphogenetic movements in the neural plate of the newt Taricha torosa. Dev Biol 18:537–552

Burton H, Loewy AD (1976) Descending projections from the marginal cell layer and other regions of the monkey spinal cord. Brain Res 116:485–491

Cabana T, Martin GF (1982) The origin of brainstem spinal projections at different stages of development in the North-American opossum. Dev Brain Res 2:163–168

Cabana T, Martin GF (1984) Developmental sequence in the origin of descending spinal pathways. Studies using retrograde transport techniques in the North-American opossum (Didelphis virginiana). Dev Brain Res 15:247–263

Cabana T, Martin GF (1985) Corticospinal development in the North-American opossum: evidence for a sequence in the growth of cortical axons in the spinal cord and for transient projections. Dev Brain Res 23:69–80

Cabana T, Martin GF (1986) The adult organization and development of the rubrospinal tract. An experimental study using the orthograde transport of WGA-HRP in the North-American opossum. Dev Brain Res 30:1–11

Cabral F, Gottesman MM (1979) Phosphorylation of the 10-Nm filament protein from Chinese hamster ovary cells. J Biol Chem 254:6203–6206

Caceras A, Binder LI, Payne MR, Bender P, Rebhun L, Steward O (1984) Differential subcellular localization of tubulin and the microtubule-associated protein MAP 2 in brain tissue as revealed by immunocytochemistry with monoclonal hybridoma antibodies. J Neurosci 4:394–410

Caffe AR, van Leeuwen FW (1983) Vasopressin-immunoreactive cells in the dorsomedial hypothalamic region, medial amygdaloid nucleus and locus coeruleus in the rat. Cell Tissue Res 233:23–33

Carden MJ, Schlaepfer WW, Lee VMY (1985) The structure, biochemical properties and immunogenicity of neurofilament peripheral regions as determined by phosphorylation. J Biol Chem 260:9805–9817

Caroni P, Schwab ME (1989) Codistribution of neurite growth inhibitors and oligodendrocytes in rat CNS: appearance follows nerve fiber growth and precedes myelination. Dev Biol 136(2):287–298

Caughell KA, Flumerfelt BA (1977) The organization of the cerebellorubral projection: an experimental study in the rat. J Comp Neurol 176:295–306

Chamak B, Prochiantz A (1989) Influence of extracellular matrix proteins on the expression of neuronal polarity. Development 106(3):483–491

Chandebois R (1976) Histogenesis and morphogenesis in planarian regeneraton. In: Wolsky A (ed) Monographs in developmental biology, vol 11. Karger, Basel

Chandebois R, Faber J (1983) Automation in animal development. In: Wolsky A (ed) Monographs in developmental biology, vol 16. Karger, Basel

Changeux JP (1987) Neuronal man. The biology of the mind. Oxford University Press, New York

Changeux JP, Danchin A (1976) Selective stabilization of developing synapses as a mechanism for the specification of neuronal networks. Nature 264:705–712

Chen KS, Stanfield BB (1987) Evidence that selective collateral elimination during postnatal development results in a restriction in the distribution of locus coeruleus neurons which project to the spinal cord in the rat. Brain Res 410:154–158

Cholley B, Wassef M, Arsénio-Nunes L, Bréhier A, Sotelo C (1989) Proximal trajectory of the brachium conjunctivum in rat fetuses and its early association with the parabrachial nucleus. A study combining in vitro HRP anterograde axonal tracing and immunocytochemistry. Dev Brain Res 45:185–202

Christian JL, McMahon JA, McMahon AP, Moon RT (1991) Xwnt-8, a Xenopus Wnt-1/int-1-related gene responsive to mesoderm-inducing growth factors, may play a role in ventral mesodermal patterning during embryogenesis. Development 111(4):1045–1055

Chu-Wang I-W, Oppenheim RW (1978a) Cell death of motoneurons in the chick embryo spinal cord. I. A light and electron microscopic study of naturally occurring and induced cell loss during development. J Comp Neurol 177:33–58

Chu-Wang I-W, Oppenheim RW (1978b) Cell death of motoneurons in the chick embryo spinal cord. II. A quantitative and qualitative analysis of degeneration in the ventral root, including evidence for axon outgrowth and limb innervation prior to cell death. J Comp Neurol 177:59–86

Chubb IW, Hodgson AJ, White GH (1980) Acetylcholinesterase hydrolyses substance P. Neuroscience 5:2065–2072

Clark FM, Proudfit HK (1991) The projection of locus coeruleus neurons to the spinal cord in the rat determined by anterograde tracing combined with immunocytochemistry. Brain Res 538:231–245

Cochard P, Paulin D (1984) Initial expression of neurofilaments and vimentin in the central and peripheral nervous system of the mouse embryo in vivo. J Neurosci 4:2080–2094

Coggeshall RE, Coulter JD, Willis WD (1974) Unmyelinated axons in the ventral roots of the cat lumbosacral enlargement. J Comp Neurol 153:39–58

Cowan WM (1978) Aspects of neural development. In: Porter R (ed) Neurophysiology III. International review of physiology, vol 17. University Park Press, Baltimore, pp 149–191

Crabtree JW (1990) Prenatal development of retinogeniculate projections in the rabbit: an HRP study. J Comp Neurol 299:75–88

Cuello AC, Del Fiacco M, Paxinos G (1978) The central and peripheral ends of the substance P containing sensory neurones in the rat trigeminal system. Brain Res 149:413–429

Dahl D (1987) Maturation of a large neurofilament protein (NF 150K) in rat postnatal development. J Neurosci Res 17:367–374

Dahl D (1988) Early and late appearance of neurofilament phosphorylated epitopes in rat nervous system: in vivo and in vitro study with monoclonal antibodies. J Neurosci Res 20:431–441

Dahl D, Bignami A (1985) Intermediate filaments in nervous tissue. In: Shay W (ed) Cell and muscle motility, vol 6. Plenum, New York, pp 75–96

Dahl D, Bignami A (1986) Neurofilament phosphorylation in development. A sign of axonal maturation. Exp Cell Res 162:220–230

Dahlström A, Fuxe K (1964) Evidence for the existence of monoamine-containing neurons in the central nervous system. I. Demonstration of monoamines in the cell bodies of brainstem neurons. Acta Physiol Scand 62 [Suppl 232]:1–55

Daniel H, Billard JM, Angaut P, Batini C (1987) The interposito-rubrospinal system. Anatomical tracing of a motor control pathway in the rat. Neurosci Res 5:87112

David G, Lories V, Heremans A, van der Schueren B, Cassiman JJ, van den Berghe H (1989) Membrane-associated chondroitin sulfate proteoglycans of human lung fibroblasts. J Cell Biol 108(3):1165–1173

Davies JA, Cook GMW, Stern CD, Keynes RJ (1990) Isolation from chick somites of a glycoprotein fraction that causes collapse of dorsal root ganglion growth cones. Neuron 2:11–20

De Beer FC, Thomeer RTWM, Lakke EAJF, Vries J de, Voogd J (1989) Spinal cord transection in fetal rats. Open intra-uterine surgery with low mortality. Brain Res 454:397–399

De Camilli P, Miller PE, Navone F, Theurkauf W, Vallee RB (1984) Distribution of microtubule-associated protein 2 in the nervous system of the rat studied by immuno-fluorescence. Neuroscience 11:817–846

Dekker JJ (1981) Anatomical evidence for direct fiber projections from the cerebellar nucleus interpositus to rubrospinal neurons. A quantitative EM study in the rat combining anterograde and retrograde intra-axonal tracing methods. Brain Res 205:229–244

Derrick GE (1937) An analysis of the early development of the chick by means of the mitotic index. J Morphol 61:257–284

Distel H, Hollander H (1980) Autoradiographic tracing of developing subcortical projections of the occipital region in fetal rabbits. J Comp Neurol 192:505–518

Dodd J, Jessel TM (1985) Lactoseries carbohydrates specify subsets of dorsal root ganglion neurons projecting to superficial dorsal horn of rat spinal cord. J Neurosci 5:3278–3294

Dodd J, Jessel TM (1986) Cell surface glycoconjugates and carbohydrate-binding proteins: possible recognition signals in sensory neuron development. J Exp Biol 124:225–238

Dodd J, Jessel TM (1988) Axon guidance and the patterning of neuronal projections in vertebrates. Science 242:692–699

Domesick VB, Morrest DK (1977) Migration and differentiation of ganglion cells in the optic tectum of the chick embryo. Neuroscience 2:459–475

Donatelle JM (1977) Growth of the corticospinal tract and the development of placing reactions in the postnatal rat. J Comp Neurol 175:207–232

Dreyer EB, Lucek P, Upchurch K, Leifert D, Lipton A (1990) An endogeneous glial receptor for the neuronal glycoprotein Thy-1: further biochemical characterization. Abstr Soc Neurosci 17(1):10.6

Düllberg S, Mai JK (1989) Age dependent changes of Leu-M1-immunoreactivity in the nucleus paraventricularis of human brains. Bull Assoc Anat (Nancy) [Suppl]:22

Eccles JC, Pritchard JJ (1937) The action potential of motoneurones. J Physiol (Lond) 89:43-45P

Eccleston PA, Silberberg DH (1985) Fibroblast growth factor is a mitogen for oligodendrocytes in vitro. Dev Brain Res 21:315–318

Edelman GM (1983) Cell adhesion molecules. Science 219:450–457

Edelman GM (1985) Molecular regulation of neural morphogenesis. In: Edelman et al. (eds) Molecular bases of neuronal development. Wiley, New York, pp 35–61

Edelman GM, Cunningham BA, Thiery JP (1990) Morphoregulatory molecules. Neurosciences Institute. Wiley, New York

Edinger L (1912) Einfürung in die Lehre von Bau und den Verrichtungen des Nervensystems. Vogel, Leipzig

Eggens I, Fenderson B, Toyokuni T, Dean B, Stroud M, Hakamori S-I (1989) Specific interaction between Lex and Lex determinants. A possible basis for cell recognition in preimplantation embryos and in embryonal carcinoma cells. J Biol Chem 264(16):9476–9484

Faúndez V, Alvarez J (1986) Microtubules and calibre in developing axons. J Comp Neurol 250:73–80

Feizi T (1985) Demonstration by monoclonal antibodies that carbohydrate structures of glycoproteins and glycolipids are onco-developmental antigens. Nature 314:53–57

Ferreira A, Busciglio J, Cáceres A (1989) Microtubule formation and neurite growth in cerebellar macroneurons which develop in vitro: evidence for the involvement of the microtubule-associated proteins, MAP 1a, HMW-MAP2 and Tau. Dev Brain Res 49:215–228

Fitzgerald M, Reynolds ML, Benowits LI (1991) Gap-43 expression in the developing rat spinal cord. Neuroscience 41:187–199

Flumerfelt BA, Gwyn DG (1973) Synaptology and afferent connections of the red nucleus in the rat. Anat Rec 175:321

Flumerfelt BA, Gwyn DG (1974) The red nucleus of the rat: its organization and interconnections. J Anat 118:374–376

Flumerfelt BA, Hrycyshyn AW (1985) Precerebellar nuclei and red nucleus. In: Paxinos G (ed) The rat nervous system. Academic, Sidney, pp 221–250

Forehand CJ (1990) Morphology of sympathetic preganglionic neurons in the neonatal rat spinal cord: an intracellular horseradish peroxidase study. J Comp Neurol 298(3):334–342

Fox N, Damjanov I, Knowles BB, Solter D (1983) Immunocytochemical localization of the mouse stage-specific embryonic antigen 1 in human tissue and tumours. Cancer Res 43:669–678

Fritschy J-M, Grzanna R (1990a) Demonstration of two separate descending noradrenergic pathways to the rat spinal cord: evidence for an intragriseal trajectory of locus coeruleus axons in the superficial layers of the dorsal horn. J Comp Neurol 291:553–582

Fritschy J-M, Grzanna R (1990b) Distribution of locus coeruleus axons within the rat brainstem demonstrated by Phaseolus vulgaris leucoagglutinin anterograde tracing in combination with dopamine-2-hydroxylase immunofluorescence. J Comp Neurol 293:616–631

Fritschy J-M, Lyons WE, MullenCA, Kosofsky CK, Molliver ME, Grzanna R (1987) Distribution of locus coeruleus axons in the rat spinal cord: a combined anterograde transport and histochemical study. Brain Res 437:176–180

Fujita S (1964) Analysis of neuron differentiation in the central nervous system by tritiated thymidine autoradiography. J Comp Neurol 122:311–328

Fulcrand J, Privat A (1977) Neuroglial reactions secondary to wallerian degeneration in the optic nerve of the postnatal rat: ultrastructural and quantitative study. J Comp Neurol 176:189–224

Gambetti P, Autilio-Gambetti L, Papasozomenos SC (1981) Bodian's silver method stains neurofilament polypeptides. Science 213:1521–1522

Gard DL, Lazarides E (1982) Cyclic AMP-modulated phosphorylation of intermediate filament proteins in cultured avian myogenic cells. Mol Cell Biol 2:1104–1114

Garner CC, Tucker RP, Matus A (1988) Selective localization of messenger RNA for cytoskeletal protein MAP 2 in dendrites. Nature 336:674–677

Carner CC, Garner A, Huber G, Kozak C, Matus A (1990) Molecular cloning of microtubule-associated protein 1 (MAP 1A) and microtubule-associated protein 5 (MAP 1B): identification of distinct genes and their differential expression in developing brain. J Neurochemistry 55(1):146–154

Gasser UE, Hatte ME (1990) Central nervous system neurons migrate on astroglial fibers from heterotopic brain regions in vitro. Proc Natl Acad Sci USA 87(12):4543–4547

Geiger B (1987) Intermediate filaments looking for a function. Nature 329:392–393

Gensburger C, Labourdette G, Sensenbrenner M (1987) Brain basic fibroblast growth factor stimulates the proliferation of rat neuronal precursor cells in vitro. FEBS Lett 217:1–5

Georgatos SD, Blodel G (1987a) Two distinct attachment sites for vimentin along the plasma membrane and the nuclear envelope in avian erythrocytes: a basis for a vectorial assembly of intermediate filaments. J Cell Biol 105:105–115

Georgatos SD, Blodel G (1987b) Laminin B constitutes an intermediate filament attachment site at the nuclear envelope. J Cell Biol 105:117–127

Georges E, Lefebvre S, Mushynski WE (1986) Dephosphorylation of neurofilaments by exogenous phosphatase has no effect on reassembly of subunits. J Neurochem 47:477–483

Giesler GJ, Cannon JT, Urca G, Liebeskind JC (1978) Long ascending projections from substantia gelatinosa Rolandi and the subjacent dorsal horn in the rat. Science 202:984–986

Giesler GJ, Ménétrey D, Basbaum AI (1979) Differential origin of spinothalamic tract projections to medial and lateral thalamus in the rat. J Comp Neurol 184:107–126

Giesler GJJ, Spiel HR, Willis WD (1981) Organization of spinothalamic tract axons within the rat spinal cord. J Comp Neurol 195:243–252

Gillette R (1944) Cell number and cell size in the ectoderm during neurulation. J Exp Zool 96:201–222

Gillilan LA (1943) The nuclear pattern of the non-tectal portions of the midbrain and isthmus in rodents. J Comp Neurol 78:213–251

Glick MC, Santer UV (1982) Expression of glycoproteins on membranes of human tumour cells. In: Akoyunoglou G (ed) Cell function and differentiation, part A. Liss, New York, pp 371–383

Glover JC (1991) Inductive events in the neural tube. TINS 14:424–428

Godement P, Vanselow J, Thanes S, Bonhoeffer F (1987) A study in developing visual systems of a new method of staining neurons and their processes in fixed tissue. Development 101:697–713

Goehring JH (1928) An experimental analysis of the motor-cell columns in the cervical enlargement of the spinal cord of the albino rat. J Comp Neurol 46:125–153

Gooi HC, Feizi T, Kapadia A, Knowless BB, Solter D, Evans MJ (1981) Stage-specific embryonic antigen involves (α)-1-3 fucosylated type II blood group chains. Nature 292:156–158

Gordon-Weeks PR, Lockerbie RO, Pearce BR (1984) Uptake and release of [^{3}H]GABA by growth cones isolated from neonatal rat. Neurosci Lett 52:205–210

Gorgels TGMF (1990) A quantitative analysis of axon outgrowth, axon loss, and myelination in the rat pyramidal tract. Dev Brain Res 54:51–61

Goslin K, Schreyer DJ, Skene JHP, Banker G (1988) Development of neuronal polarity: GAP-43 distinguishes axonal from dendritic growth cones. Nature 336:672–674

Greenfield SA (1984) Acetylcholinesterase may have novel functions in the brain. TINS 7:364–368

Greenfield SA (1991) A noncholinergic action of acetylcholinesterase (AChE) in the brain: from neuronal secretion to the generation of movement. Cell Mol Neurobiol 11(1): 55–77

Gribnau AAM, de Kort EJM, Dederen PWJC, Nieuwenhuys R (1986) On the development of the pyramidal tract in the rat. II. An antegrade tracer study on the outgrowth of the corticospinal fibres. Anat Embryol 175:101–110

Grumet M, Hoffman S, Edelman GM (1984) Two antigenetically related neuronal CAMs of different specificities mediate neuron-neuron and glia-neuron interaction. Proc Natl Acad Sci USA 82:8075–8079

Gwyn DG, Flumerfelt BA (1974) A comparison of the distribution of cortical and cerebellar afferents in the red nucleus of the rat. Brain Res 69:130–135

Gwyn DG, Waldron HA (1968) A nucleus in the rat dorsolateral funiculus of the spinal cord of the rat. Brain Res 10:342–351

Hamburger V (1975) Cell death in the development of the lateral motor column of the chick embryo. J Comp Neurol 160:535–546

Hamburger V, Brunso-Brechtold JK, Yip JW (1981) Neuronal death in the spinal ganglia of the chick embryo and its reduction by nerve growth factor. J Neurosci 1:60–71

Hancock MB, Peveto CA (1979a) A preganglionic autonomic nucleus in the dorsal gray commissure of the lumbar spinal cord of the rat. J Comp Neurol 183:65–72

Hancock MB, Peveto CA (1979b) Preganglionic neurons in the sacral spinal cord of the rat: an HRP study. Neurosci Lett 11:1–5

Hankin MH, Silver J (1986) Mechanisms of axonal guidance. The problem of intersecting fiber systems. In: Browder LW (ed) Developmental biology, vol 2. Plenum, New York, pp 565–603

Hansen SH, Stagaard M, Mollgård K (1989) Neurofilament-like pattern of reactivity in human foetal PNS and spinal cord following immunostaining with polyclonal anti-glialfibrillary acidic protin antibodies. J Neurocytol 18:427–436

Hardy R, Reynolds R (1991) Proliferation and differentiation potential of rat forebrain oligodendroglial progenitors both in vitro and in vivo. Development 111(4):1061–1080

Hebel R, Stromberg MW (1986) Anatomy and embryology of the laboratory rat. Biomed Verlag, Würthsee

Heffner CD, Lumsden AGS, O'Leary DDM (1990) Target control of collateral extension and directional axon growth in the mammalian brain. Science 247:217–220

Held H (1890) Der Ursprung des tiefen Markes der Vierhugelregion. Neurol Zentralbl 481–483

Hicks SP, D'Amato CJ (1963) Malformation and regeneration of the mammalian retina following experimental radiation. In: Michaux L, Field M (eds) Les phakomatoses cérébrales; deuxième colloque international malformations congénitales de l'encéphale. SPEI, Paris, pp 45–51

Hicks SP, D'Amato CJ (1966) Effects of ionizing radiations on mammalian development. In: Woollam DHM (ed) Advances in teratology. Logos, London, pp 196–250

Hicks SP, D'Amato CJ, Lowe MJ (1959) The development of the mammalian nervous system. I. Malformations of the brain, especially the cerebral cortex, induced in rats by radiation. II. Some mechanisms of the malformations of the cortex. J Comp Neurol 113:435–469

Hicks SP, D'Amato CJ, Coy MC, O'Brien ED, Thurston JM, Joftes DL (1961) Migrating cells in the developing central nervous system studied by their radiosensitivity and tritiated thymidine uptake. In: Fundamental aspects of radiosensitivity. Brookhaven Symposium in Biology, pp 246–261

Hinrichsen CFL, Watson CD (1983) Brainstem projections to the facial nucleus of the rat. Brain Behav Evol 22:153–163

Hirano M, Dembitzer HM (1978) Morphology of normal central myelinated axons. In: Waxman SG (ed) Physiology and pathobiology of axons. Raven, New York, pp 65–82

Hirano M, Goldman JE (1988) Gliogenesis in rat spinal cord: evidence for origin of astrocytes and oligodendrocytes from radial precursors. J Neurosci Res 21:155–167

Hirokawa N, Glicksman MA, Willard MB (1984) Organization of mammalian neurofilament polypeptides within the neural skeleton. J Cell Biol 98:1523–1536

His W (1889) Die Neuroblasten und deren Entstehung in embryonalen Mark. Hirzel, Leipzig

His W (1892a) Zur allgemein Morphologie des Gehirns. Arch Anat Entwicklungsgesch 1892:346–383

His W (1982b) Zur Nomenclatur des Gehirns und Rückenmarkes. Arch Anat Entwicklungsgesch 1892:425–428

Hökfelt T, Elde R, Johansson O, Luft R, Nilsson G, Arimura A (1976) Immunohistochemical evidence for separate populations of somatostatin-containing and substance P-containing primary afferent neurons. Neuroscience 1:131–136

Holets VR, Hökfelt T, Rökaeus A, Terenius L, Goldstein M (1988) Locus coeruleus neurons in the rat containing neuropeptide Y, tyrosine hydroxylase or galanin and their efferent projections to the spinal cord, cerebral cortex and hypothalamus. Neuroscience 24:893–906

Hollyday M (1985) Development of motor innervation in vertebrate limbs. In: Edelman GM et al. (eds) Molecular bases of neuronal development. Wiley, New York, pp 243–265

Holstege G, Griffiths D (1990) Neuronal organization of micturition. In: Paxinos G (ed) The human nervous system. Academic, Sidney, pp 297–305

Holstege G, Tan J (1988) Projections from the red nucleus and surrounding areas to the brainstem and spinal cord in the cat. An HRP and autoradiographical tracing study. Behav Brain Res 28:33–57

Holstege G, Griffiths D, de Wall H, Dalm E (1986) Anatomical and physiological observations on supraspinal control of bladder and urethral sphincter muscles in the cat. J Comp Neurol 250:449–461

Holstege G, Blok BF, Ralston DD (1988) Anatomical evidence for red nucleus projections to motoneuronal cell groups in the spinal cord of the monkey. Neurosci Lett 95:97–101

Holtfreter J (1944) Significance of the cell membrane in embryonic processes. Ann N Y Acad Sci 49:709–760

Holzman E, Peterson ER (1969) Uptake of protein by mammalian neurons. J Cell Biol 40:863–869

Hommes OR, Leblond CP (1967) Mitotic division of neuroglia in the normal adult rat. J Comp Neurol 129:269–278

Horder TJ, Martin KAC (1979) Morphogenetics as an alternative to chemospecificity in the formation of nerve connections. Symp Soc Exp Biol 32:275–359

Horwitz A, Duggan K, Buck C, Beckerle M, Burridge K (1986) Interaction of plasma membrane fibronectin receptor with talin-a transmembrane linkage. Nature 320:531–533

Houle J, Federoff S (1983) Temporal relationship between appearance of vimentin and neural tube development. Dev Brain Res 9:189–195

Huber G, Matus A (1984) Differences in the cellular distribution of two microtubule-associated proteins, MAP 1 and MAP 2 in the rat brain. J Neurosci 4:151–160

Huisman AM, Kuypers HGJM, Verburgh CA (1981) Quantitative differences in collateralization of the descending spinal pathways from red nucleus and other brain stem cell groups in rat as demonstrated with the multiple fluorescent retrograde tracer technique. Brain Res 209:271–286

Hutchins JB, Casagrande VA (1988) Development of acetylcholinesterase activity in the lateral geniculate nucleus. J Comp Neurol 275:241–253

Hüther G, Luppa H (1979) The multiple forms of brain acetylcholinesterase. III. Implications for the histochemical demonstration of acetylcholinesterase. Histochemistry 63:115–121

Hüther G, Luppa H, Ott T (1978) The multiple forms of brain acetylcholinesterase. II. A suggestion of their functional importance. Histochemistry 55:55–62

Jacobson M (1979) Developmental neurobiology. Plenum, New York

Jessel TM (1988) Adhesion molecules and the hierarchy of neural development. Neuron 1:3–13

Jessel TM, Dodd J (1985) Structure and expression of differentiation antigens on functional subclasses of primary sensory neurons. Philos Trans R Soc [B] 308:271–281

Jones BE, Yang T-Z (1985) The efferent projection from the reticular formation and the locus coeruleus studied by anterograde and retrograde axonal transport in the rat. J Comp Neurol 242:56–92

Joosten EAJ, Gribnau AAM (1989) Astrocytes and guidance of outgrowing corticospinal tract axons in the rat. An immunocytochemical study using anti-vimentin and anti-glial fibrillary acidic protein. Neuroscience 31(2):439–452

Joosten EAJ, van Eden CG (1989) An anterograde tracer study on the development of corticospinal projections from the medial prefrontal cortex in the rat. Dev Brain Res 45:313–319

Joosten EAJ, Gribnau AAM, Dederen PWJC (1987) An anterograde tracer study of the developing corticospinal tract in the rat: three components. Dev Brain Res 36:121–130

Joosten EAJ, van der Ven PFM, Hooiveld MHW, ten Donkelaar HJ (1991) Induction of corticospinal target finding by release of a diffusible, chemotropic factor in cervical grey matter. Neurosci Lett 128:25–28

Kanemitsu A (1971) Relation entre la taille des neurones et leur époque d'apparition dans la moelle épiniere chez le poulet. Etude autoradiographique et caryométrique. Proc Jpn Acad 47:432–437

Kannagi R, Levery SB, Ishigami F, Hakomori SI, Shevinsky LH, Knowles BB, Solter D (1983) New globoseries glycosphingolipids in human teratocarcinoma reactive with the monoclonal antibody directed to a developmentally regulated antigen, stage-specific embryonic antigen 3. J Biol Chem 258:8934–8942

Karfunkel P (1974) The mechanisms of neural tube formation. Int Rev Cytol 38:245–271

Karnovsky MJ, Roots N (1964) A "direct colouring" thiocholine method for cholinesterases. J Histochem Cytochem 12:219–221

Keizer K, Kuypers HGJM (1989) Distribution of corticospinal neurons with collaterals to the lower brain stem reticular formation in monkey (Macaca fascicularis). Exp Brain Res 74:311–318

Keizer K, Kuypers HGJM, Huisman AM, Dann O (1983) Diamidino yellow hydrochloride (DY.2HCl); a new fluorescent retrograde neuronal tracer, which migrates only very slowly out of the cell and can be used in conjunction with TB and FB in double labeling experiments. Exp Brain Res 51:179–191

Kennedy PR (1987) Double labeling of red nucleus neurons from dye injections into the inferior olive and dorso-lateral funiculus of the spinal cord in rat. Soc Neurosci Abstr 13:852

Kerrebroek, Debruyne (eds) Dysfunction of the lower urinary tract. Present achievements and future perspectives. Medicom Europe, Bussum, pp 3–9

Kidd GJ, Hauer PE, Trapp BD (1990) Axons modelate myelin protein messenger RNA levels during central nervous system myelination in vivo. J Neurosci Res 26(4):409–418

Koelle GB (1954) The histochemical localization of cholinesterases in the central nervous system of the rat. J Comp Neurol 100:211–235

Kort EJM de, Gribnau AAM, van Aanholt HTH, Nieuwenhuys R (1985) On the development of the pyramidal tract in the rat. I. The morphology of the growth zone. Anat Embryol (Berl) 172:195–204

Kostovic I, Rakic P (1984) Development of prestriate visual projections in the monkey and human fetal cerebrum revealed by transient cholinesterase staining. J Neurosci 4:25–42

Kroger S, Walter J (1991) Molecular mechanisms separating two axonal pathways during embryonic development of the avian tectum. Neuron 6(2):291–303

Kruse MN, DeGroat WC (1990) Micturition reflexes in decerebrate and spinalized neonatal rats. Am J Physiol 258 (Regulatory Integrative Comp Physiol 27): R1508–R1511

Kruse MN, Noto H, Roppolo JR, DeGroat WC (1990) Pontine control of the urinary bladder and external urethral sphincter in the rat. Brain Res 532:182–190

Kudo N, Yamada T (1987) Morphological and physiological studies of development of the monosynaptic reflex pathway in the rat lumbar spinal cord. J Physiol (Lond) 389:441–459

Kuhlenbeck H (1975) The central nervous system of vertebrates, vol 4. Spinal cord and deuterencephalon. Karger, Basel

Kuraishi Y, Hirota N, Sato Y, Hin Y, Satoh M, Takagi H (1985) Evidence that substance P and somatostatin transmit separate information related to pain in the spinal dorsal horn. Brain Res 325:294–298

Kuwada JY, Bernhardt RR (1990) Axonal outgrowth by identified neurons in the spinal cord of zebrafish embryos. Exp Neurol 109(1):29–34

Kuypers HGJM (1981) Anatomy of the descending pathways. In: Brooks VB (ed) Handbook of physiology. I. The nervous system, vol 2. Motor control, part I. American Physiological Society, Bethesda, pp 597–666

Kuypers HGJM, Huisman AM (1984) Fluorescent neuronal tracers. In: Fedoroff S (ed) Advances in cellular neurobiology, vol 5: labelling methods applicable to the study of neuronal pathways. Academic, London, pp 307–340

Kuznetsov SA, Rodinov VI, Gelfand GI, Rosenblat VA (1981) Microtubule-associated protein MAP 1 promotes microtubule assembly in vitro. FEBS Lett 135:241–244

Lakke EAJF, Hinderink JB (1989) Development of the spinal projections of the nucleus paraventricularis hypothalami of the rat: an intra-uterine WGA-HRP study. Dev Brain Res 49:115–121

Lakke EAJF, Marani E (1991) The prenatal descent of rubro-spinal fibers through the spinal cord of the rat. J Comp Neurol 314:67–78

Lakke EAJF, Hinderink JB, Marani E (1990) The development of the descending projections of the nucleus tegmentalis laterodorsalis in the rat. In: van Kerrebroek PEV, Debruyne FMJ (eds) Dysfunction of the lower urinary tract. Present achievements and future perspectives. Medicom Europe BV, Bussum, the Netherlands, pp 3–9

Lamotte C (1977) Distribution of the tract of Lissauer and the dorsal root fibers in the primate spinal cord. J Comp Neurol 172:529–562

Lander AD, Fuji DK, Reichardt LF (1985) Purification of a factor that promotes neurite outgrowth: isolation of laminin and associated molecules. J Cell Biol 101:898–913

Landis DMD, Reese TS (1983) Cytoplasmic organization in cerebellar dendritic spines. J Cell Biol 97:1169–1178

Lauder JM, Bloom FE (1974) Ontogeny of monoamine neurons in the locus coeruleus, raphe nuclei and substantia nigra of the rat. I. Cell differentiation. J Comp Neurol 155: 469–482

LaVail JH, LaVail MM (1974) The retrograde intra-axonal transport of horseradish peroxidase in the chick visual system: a light and electron microscopic study. J Comp Neurol 157:303–358

Layer PG (1983) Comparative localization of acetylcholinesterase and pseudocholinesterase during morphogenesis of the chick brain. Proc Natl Acad Sci USA 80:6413–6417

Layer PG, Sporns O (1987) Spatiotemporal relationship of embryonic cholinesterases with cell proliferation in chicken brain and eye. Proc Natl Acad Sci USA 84:284–288

Layer PG, Rommel S, Balthoff H, Hengstenberg R (1988) Independent spatial waves of biochemical differentiation along the surface of chicken brain as revealed by the sequential expression of acetylcholinesterase. Cell Tissue Res 251:587–595

Lazarides E (1980) Intermediate filaments as mechanical integrators of cellular space. Nature 283:249–256

Lazarides E (1982) Intermediate filaments: a chemically heterogeneous, developmentally regulated class of proteins. Annu Rev Biochem 51:219–250

Leber SM, Sanes JR (1990) Lineage analysis with a recombinant retrovirus: application to chick spinal motor neurons. Experientia 46(9):929–940

Lee VMY, Page C (1984) The dynamics of nerve growth factor-induced neurofilament and vimentin filament expression and organization in PC12 cells. J Neurosci 4:1705–1714

Lemmon V, Burden SM, Payne HR, Elmslie GJ, Hlavin ML (1992) Neurite growth on different substrates: permissive versus instructive influences and the role of adhesive strenght. J Neurosci 12(3):818–826

Leong SK (1983) Localizing the corticospinal neurons in neonatal, developing and mature albino rat. Brain Res 265:1–10

Leong SK, Shieh JY, Wong WC (1984) Localizing spinal-cord-projection neurons in neonatal and immature albino rat. J Comp Neurol 228:18–23

Light AR, Perl ER (1979) Reexamination of the dorsal root projection to the spinal dorsal horn including observations on the differential termination of coarse and fine fibres. J Comp Neurol 186:117–13

Lipton SA (1989) Growth factors for neuronal survival and process regeneration. Arch Neurol 46:1241–1248

Liu RPC (1983) Laminar origins of spinal projection neurons to the periaquaductal gray of the rat. Brain Res 264:118–122

Llinas R, Sugimori M (1979) Calcium conductances in Purkinje cell dendrites: their role in development and integration. Prog Brain Res 51:323–334

Lockerbie RO, Gordon-Weeks PR (1985) 3-Aminobutyric acid-A (GABA A) receptors modulate [^{3}H]GABA release from isolated neuronal growth cones in the rat. Neurosci Lett 55:273–277

Lockridge O (1982) Substance P hydrolysis by human serum cholinesterase. J Neurochem 39:106–110

Loewy AD, Saper CB, Baker RP (1979) Descending projections from the pontine micturition center. Brain Res 172:533–538

Ludwin SK, Kosek JC, Eng LF (1976) The topographical distribution of S-100 and GFA proteins in the adult rat brain: an immunohistochemical study using horseradish peroxidase labelled antibodies. J Comp Neurol 165:197–208

Luiten PGM, ter Horst GJ, Karst H, Steffens AB (1985) The course of paraventricular hypothalamic efferents to autonomic structures in medulla and spinal cord. Brain Res 329:374–378

Mai JK, Reifenberger G (1988) Distribution of the carbohydrate epitope 3-fucosyl-N-acetyl-lactosamine (FAL) in the adult human brain. J Chem Neuroanat 1(5):255–285

Maissonpierre PC, Belluscio L, Friedman B, Alderson RF, Wiegand SJ, Furth ME, Lindsay RM, Yancopoulos GD (1990) NT-3, BDNF, and NGF in the developing rat nervous system: parallel as well as reciprocal patterns of expression. Neuron 5:501–509

Marani E (1978) A method for orientating cryostat sections for three-dimensional reconstructions. Stain Technol 53:265–268

Marani E (1981) Enzyme histochemistry. In: Lahue R (ed) Methods in neurobiology, vol I. Plenum, New York, pp 481–583

Marani E (1986) Topographic histochemistry of the cerebellum. Prog Histochem Cytochem 16(4):110–116

Marani E, Mai JK (1992) Expression of the carbohydrate epitope 3-fucosyl-N-acetyl-lactosamine (CD15) in the vertebrate cerebellar cortex. Histochem J 24:852–868

Marani E, Tetteroo PAT (1983) A longitudinal bandpattern for the monoclonal human granulocyte antibody B4,3 in the cerebellar external layer of the immature rabbit. Histochemistry 78:157–161

Marani E, Voogd J (1977) An acetylcholinesterase band-pattern in the molecular layer of the cat cerebellum. J Anat 124:335–345

Marani E, Voogd J, Boekee A (1977) Acetylcholinesterase staining in subdivisions of the cat's inferior olive. J Comp Neurol 174:209–226

Marani E, Epema A, Brown B, Tetteroo P, Voogd J (1986a) The development of longitudinal patterns in the rabbit cerebellum. Acta Histochem [Suppl] XXXII:S53–S58

Marani E, van Oers JWAM, Tetteroo PAT, Poelmann RE, van Veeken JGPM, Deenen MGM (1986b) Stage specific embryonic carbohydrate surface antigens of primordial germ cells in mouse embryos: FAL (SSEA-1) and globoside (SSEA-3). Acta Morphol Neerl Scand 24:103–110

Marani E, Deenen M, Maassen JA (1992) The expression of CD15 in dissociated cultured rat dorsal ganglion cells. Histochem J 24:833–841

Martin GF, Beattie MS, Bresnahan JC, Henkel CK, Hughes HC (1975) Cortical and brainstem projections to the spinal cord of the American opossum (Didelphis marsupialis virginiana). Brain Behav Evol 12:270–310

Martin GF, Beals JK, Culberson JL, Dom R, Goode G, Humbertson AO (1978) Observations on the development of brainstem-spinal systems in the North-American opossum. J Comp Neurol 181:271–289

Martin GF, Cabana T, Culberson TL, Curry JJ, Tschismadia I (1980) The early development of corticobulbar and corticospinal systems: studies using the North-American opossum. Anta Embryol 161:197–213

Martin GF, Cabana T, Humberston AO Jr (1981) Evidence for a lack of distinct rubrospinal somatotopy in the North American opossum and for collateral innervation of the cervical and lumbar enlargements by single rubral neurons. J Comp Neurol 201:255–263

Martin GF, Cabana T, DiTirro FJ, Ho RH, Humbertson AO (1982) The development of descending fiber connections. Studies using the North-American oppossum. In: Kuypers HGJM, Martin GF (eds) Descending pathways to the spinal cord. Prog Brain Res 57:131–141

Martin GF, Culberson JL, Hazlett JC (1983) Observations on the early development of ascending spinal pathways. Studies using the North-American opossum. Anat Embryol (Berl) 166:191–207

Martin GF, Cabana T, Hazlett JC (1986) The development of rubrospinal, cerebellorubral, and corticorubral connections in the North-American opossum. Evidence for asynchronism. Neurochem Pathol 5:221–236

Martin GF, Cabana T, Hazlett JC (1988) The development of selected rubral connections in the North-American opossum. Behav Brain Res 28:21–28

Martinou JC, Le Van Thai A, Cassar G, Roubinet F, Weber MJ (1989) Characterization of two factors enhancing choline acetyltransferase activity in cultures of purified rat motoneurons. J Neurosci 9:3645–3656

Mattews MA, Duncan D (1971) A quantitative study of morphological changes accompanying the initiation and progress of myelin production in the dorsal funiculus of the rat spinal cord. J Comp Neurol 142:1–22

Matus A (1988) Microtubule-associated proteins: their potential role in determining neuronal morphology. Annu Rev Neurosci 11:29–44

McClung JR, Castro AC (1978) Rexed's laminar scheme as it applies to the rat cervical spinal cord. Exp Neurol 58:145–148

McConnel SK (1988) Development and discision-making in the mammalian cerebral cortex. Brain Res 472:1–23

McGeer EG, McGeer PL (1989) Effect of fixatives on rat brain acetylcholinesterase activity. J Neurosci Methods 28:235–237

Menétrey D, Chaouch A, Binder D, Besson JM (1982) The origin of the spinomesencephalic tract in the rat: an anatomical study using the retrograde transport of horseradish peroxidase. J Comp Neurol 206:193–207

Mesulam MM (1982) Tracing neural connections with horseradish peroxidase. Wiley, New York

Miller MW (1987) The origin of corticospinal projection neurons in rat. Exp Brain Res 67:339–351

Miller MW (1988) Development of projection and local circuit neurons in neocortex. In: Peters A, Jones EG (eds) Cerebral cortex, vol 7. Development and maturation of cerebral cortex. Plenum, New York, pp 133–175

Molander C, Xu Q, Grant G (1984) The cytoarchitectonic organization of the spinal cord in the rat. I. The lower thoracic and lumbosacral cord. J Comp Neurol 280:133–141

Molenaar I, Kuypers HGJM (1978) Cells of origin of propriospinal fibres and of fibres ascending to suprasapinal levels. A HRP study in cat and rhesus monkey. Brain Res 152:429–450

Morrel JI, Pfaff DW (1983) Retrograde HRP identification of neurons in the rhombencephalon and spinal cord of the rat that project to the dorsal mesencephalon. Am J Anat 167:229–240

Morrison RS, Sharma A, De Vellis J, Bradshaw RA (1986) Basic fibroblast growth factor supports the survival of cerebral cortical neurons in primary culture. Proc Natl Acad Sci USA 83:7537–7541

Moscona AA (1952) Cell suspensions from organ rudiments of chick embryos. Exp Cell Res 3:536–539

Murray HM, Gurule ME (1979) Origin of the rubrospinal tract in the rat. Neurosci Lett 14:19–25

Nadelhaft I, DeGroat WC, Morgan C (1980) Location and morphology of parasympathetic preganglionic neurons in the sacral spinal crod of the cat revealed by retrograde transport of horseradish peroxidase. J Comp Neurol 193:265–283

Nauta HJW, Pritz MB, Lasek RJ (1974) Afferents to the rat caudoputamen studied with horseradish peroxidase. An evaluation of a retrograde neuroanatomical research method. Brain Res 67:219–238

Niedieck B, Löhler J (1987) Expression of 3-fucosyl N-acetyl lactosamine on glia cells and its putative role in cell adhesion. Acta Neuropathol (Berl) 75:173–184

Nixon RA, Shea TB (1992) Dynamics of neuronal intermediate filaments: a developmental perspective. Cell Motil Cytoskeleton 22:81–91

Nixon AR, Sihag RK (1992) Neurofilament phophorylation: a new look at regulation and function. TINS 14(11):501–506

Nixon RA, Lewis SE, Marotta CA (1987) Posttranslational modification of neurofilament proteins by phosphate during axoplasmic transport in retinal ganglion cell neurons. J Neurosci 7:1145–1158

Noble M, Murray K (1984) Purified astrocytes promote the in vitro division of a bipotential glial progenitor cell. EMBO J 3:2243–2247

Nordlander RH, Baden ST, Ryba TMJ (1985) Development of early brainstem projections to the tail spinal cord of Xenopus. J Comp Neurol 231:519–529

Nornes HO, Das GD (1974) Temporal pattern of neurogenesis in spinal cord of rat. I. An autoradiographic study – times and sites of origin and migration and settling patterns of neuroblasts. Brain Res 73:121–138

Nornes HO, Hart H, Carry M (1980a) Pattern of development of ascending and descending fibres in embryonic spinal cord of chick. I. Role of position information. J Comp Neurol 192:119–132

Nornes HO, Hart H, Carry M (1980b) Pattern of development of ascending and descending fibres in embryonic spinal cord of chick. II. A correlation with behaviourial studies. J Comp Neurol 192:133–14

Nyberg-Hansen R, Brodal A (1963) Sites and mode of termination of rubrospinal fibres in the cat. An experimental study with silver impregnation methods. J Anat 98: 235–253

Oblinger MM (1987) Characterization of posttranslational processing of the mammalian high molecular weight neurofilament protein in vivo. J Neurosci 7:2510–2521

Okado N, Oppenheim RW (1985) The onset and development of descending pathways to the spinal cord in the chick embryo. J Comp Neurol 232:143–161

Okado N, Yoshimoto M, Furber SE (1987) Pathway formation and the terminal distribution pattern of the spinocerebellar projection in the chick embryo. Anat Embryol (Berl) 176:165–174

Oppenheim RW, Chu-Wang I-W, Maderdrut JL (1878) Cell death of motoneurons in the chick embryo spinal cord. III. The differentiation of motoneurons prior to their induced degeneration following limb-bud removal. J Comp Neurol 177:87–112

Oscarsson O (1973) Functional organization of spinocerebellar paths. In: Iggo A (ed) Somatosensory system. Springer, Berlin Heidelberg New York, pp 340–380 (Handbook of sensory physiology, vol II)

Oudega M (1990) Development of the rat spinal cord. Histochemistry of some functional and structural parameters. Thesis, Leiden, the Netherlands

Oudega M, Marani E (1990) Acetylcholinesterase in the developing rat spinal cord: an enzyme histochemical study. Eur J Morphol 28(2–4):347–378

Oudega M, Marani E (1991) Expression of vimentin and glial fibrillary acidic protein in the developing rat spinal cord: an immunocytochemical study of the spinal cord glial system. J Anat 179:97–114

Oudega M, Riederer BM, Deenen MGM, Marani E (1990a) Expression of microtubule-associated protein 1A, 2 and 5 in the developing rat spinal cord. In: Oudega M (ed) Development of the rat spinal cord. Histochemistry of some functional and structural parameters. Thesis, Leiden, the Netherlands

Oudega M, Voormolen JHC, Marani E, Thomeer RTWM (1990b) Immunocytochemical localization of neurofilaments in the fibre systems of the developing rat spinal cord white matter. In: Oudega M (ed) Development of the rat spinal cord. Histochemistry of some functional and structural parameters. Thesis, Leiden, the Netherlands

Oudega M, Marani E, Thomeer RTWM (1992) Transient expression of stage-specific embryonic antigen-1 (CD 15) in the developing rat spinal cord. Histochem J 24:869–877

Patterson JT, Head PA, McNeill DL, Chung K, Coggeshal RE (1989) Ascending unmyelinated primary afferent fibers in the dorsal funuculus. J Comp Neurol 290: 384–390

Paxinos G, Törk I, Valentino K, Tecott L (1990) Atlas of the developing rat brain. Academic, San Diego

Petras JM, Cummings JF (1972) Autonomic neurons in the spinal cord of the rhesus monkey: a correlation of the finding of cytoarchitectonics and symphatectomy with fiber degeneration following dorsal rhizotomy. J Comp Neurol 146:184–218

Pettman B, Weibel M, Sensenbrenner M, Labourdette G (1985) Purification of two astroglial growth factors from bovine brain. FEBS Lett 189:102–108

Phelps PE, Barber RP, Vaugn JE (1991) Embryonic development of choline acetyltransferase in thoracic spinal motor neurons: somatic and autonomic neurons may be derived from a common cellular group. J Comp Neurol 307:77–86

Pippenger MA, Sims TJ, Gilmore SA (1990) Development of the rat corticospinal tract through an altered glial environment. Dev Brain Res 55:43–50

Pixley SKR, De Vellis J (1984) Transition between immature radial glia and mature astrocytes studied with a monoclonal antibody to vimentin. Dev Brain Res 15:201–209

Placzek M, Tessier-Lavigne M, Jessel TM, Dodd J (1990) Orientation of commissural axons in vitro in response to a floor plate-derived chemoattractant. Development 110:19–30

Plank J, Mai JK (1992) Developmental expression of the 3-fucosyl-N-acetyl-lactosamine/CD15 epitope by an olfactory receptor cell subpopulation and in the olfactory bulb of the rat. Dev Brain Res 66:257–261

Poltorak M, Sadoul R, Keilhauer G, Landa C, Fahrig T, Schachner M (1987) Myelin-associated glycoprotein, a member of the L2/HNK-1 family of neural cell adhesion molecules, is involved in neuron-oligodendrocyte and oligodendrocyte-oligodendrocyte interaction. J Cell Biol 105(4):1893–1899

Pompeiano O, Brodal A (1957) Experimental demonstration of a somatotopical origin of rubrospinal fibers in the cat. J Comp Neurol 108:225–252

Purves D, Hume RI (1981) The relation of postsynaptic geometry to the number of pre-synaptic axons that innervate autonomic ganglion cells. J Neurosci 1:441–452

Purves D, Lichtman JW (1980) Elimination of synaptic connections in the developing mammalian nervous system. Science 210:153–157

Purves D, Lichtman JW (1985) Principles of neuronal development. Sinauer, Massachusetts, pp 25–105

Raff MC, Miller RH, Noble M (1983) A glial progenitor cell that develops in vitro into an astrocyte or an oligodendrocyte depending on culture medium. Nature 303:390–396

Raff MC, Abney ER, Miller RH (1984) Two glial cell lineages diverge prenatally in rat optic nerve. Dev Biol 106:53–60

Rajaofetra N, Sandillon F, Geffard M, Privat A (1989) Pre- and post-natal ontogeny of serotoninergic projections to the rat spinal cord. J Neurosci Res 22:305–321

Rakic P (1972) Mode of cell migration to the superficial layers of fetal monkey neocortex. J Comp Neurol 145:61–84

Rakic P (1985) Mechanisms of neuronal migration in developing cerebellar cortex. In: Edelmen GM et al. (eds) Molecular bases of neuronal development. Wiley, New York, pp 139–161

Ralston HJ (1979) The fine structure of laminae I, II and III of the macaque spinal cord. J Comp Neurol 184:619–642

Ramon Y, Cajal S (1890) Sur l'origine et les ramifications des fibres nerveuses de la moelle embryonnaire. Anat Anz 5:85–95

Ramon Y, Cajal S (1892) The structure of the retina. (Translation 1972.) Thomas, Springfield

Ramon Y, Cajal S (1909) Histologie du systeme nerveux de l'homme et des vertébrés. I. Maloins, Paris

Rando TA, Bowers CW, Zigmond RE (1981) Localization of neurons in the rat spinal cord which project to the superior cervical ganglion. J Comp Neurol 196:73–83

Raper JA, Kapfhammer JP (1990) The enrichment of a neuronal growth cone collapsing activity from embryonic chick brain. Neuron 2:21–29

Rastan S, Thorpe SJ, Scudder P, Brown S, Gooi HC, Feizi T (1985) Cell interactions in pre-implantation embryos: evidence for involvement of saccharides of the poly-N-acetyl-lactosamine series. J Embryol Exp Morphol 87:115–128

Rauschecker JP, Singer W (1981) The effects of early visual experience on the cat's visual cortex and their possible explanation by Hebb synapses. J Physiol (Lond) 310:215–239

Reid JM, Gwyn DG, Flumerfelt BA (1975) A cytoarchitectonic and Golgi study of the red nucleus in the rat. J Comp Neurol 162:337–362

Renshaw M (1940) Activity in the simplest spinal reflex pathways. J Neurophysiol 3:373–387

Réthelyi M (1977) Preterminal and terminal arborizations in the substantia gelatinosa of the cat's spinal cord. J Comp Neurol 172:511–528

Retzius G (1898) Zur Frage von der Endigungsweise der peripherischen sensibelen Nerven. Biol Untersuch 8:114

Rexed B (1954) A cytoarchitectonic atlas of the spinal cord in the cat. J Comp Neurol 100:297–379

Riederer BM (1989) Antigen preservation tests for immunocytochemical detection of cyto-skeletal proteins: influence of aldehyde fixatives. J Histochem Cytochem 37:675–681

Riederer BM (1990) Some aspects of the neuronal cytoskeleton in development. Eur J Morphol 28(2–4):347–378

Riederer BM, Matus A (1985) Differential expression of distinct microtubule-associated proteins during brain development. Proc Natl Acad Sci USA 82:6006–6009

Riederer BM, Cohen R, Matus A (1986) MAP5: a novel brain microtubule-associated protein under strong developmental regulation. J Neurocytol 15:763–775

Riederer BM, Guadano-Ferraz A, Innocenti I (1991) Difference in distribution of micro-tubule-associated proteins 5a and 5b during the development of cerebral cortex and corpus callosum in cats: dependence on phosphorylation. Dev Brain Res 56(2):235–243

Rienitz A, Grenningloh G, Hermans-Borgmeyer I, Kirsch J, Littauer UZ, Prior P, Gundelfinger ED, Schmitt B, Betz H (1989) Neuraxin, a novel putative structural protein of the rat central nervous system that is immunologically related to microtubule-associated protein 5. EMBO J 8:2879–2888

Rogers RC (1985) An inexpensive picoliter-volume pressure ejection system. Brain Res Bull 15:669–671

Rozeik C, Von Keyserlingk D (1987) The sequence of myelination in the brainstem of the rat monitored by basic myelin protein immunohistochemistry. Dev Brain Res 35:183–190

Ruggiero DA, Ross CA, Reis DJ (1981) Projections from the spinal trigeminal nucleus to the entire length of the spinal cord in the rat. Brain Res 225:225–233

Ruit KG, Elliot JL, Osborne PA, Yan Q, Snider WD (1992) Selective dependence of mammalian dorsal root ganglion neurons on nerve growth factor during embryonic development. Neuron 8:573–587

Rutishauser U, Gall WE, Edelman GM (1978) Adhesion among neural cells in the chick embryo. I. Role of the cell surface molecule CAM in the formation of neurite bundles in cultures of spinal ganglia. J Cell Biol 79:382–393

Rye DB, Lee HJ, Saper CB, Wainer BH (1988) Medullary and spinal efferents of the pedunculopontine tegmental nucleus and adjacent mesopontine tegmentum in the rat. J Comp Neurol 269:315–341

Saper CB, Loewy AD (1980) Efferent connections of the parabrachial nucleus of the rat. Brain Res 197:291–317

Satoh KS, Tohyama M, Yamamoto K, Sakumoto T, Shimizu N (1977) Noradrenaline innervation of the spinal cord studied by the horseradish peroxidase method combined with monoamine oxidase staining. Exp Brain Res 30:175–186

Satoh KS, Shimizu N, Tohyama M, Maeda T (1978a) Localization of the micturition reflex center at dorsolateral pontine tegmentum of the rat. Neurosci Lett 8:27–33

Satoh KS, Tohyama M, Sakumoto S, Yamamoto K, Shimizu N (1978b) Descending projection of the nucleus tegmentalis laterodorsalis to the spinal cord: studied by the horseradish peroxidase method following 6-hydroxy-dopa administration. Neurosci Lett 8:9–15

Sauer FC (1935) Mitosis in the neural tube. J Comp Neurol 62:377–406

Sauer ME (1959) Radioautographic study of the localization of newly synthesized deoxyribonucleic acid in the neural tube of the chick embryo: evidence for intermitotic migration of nuclei. Anat Rec 133:456

Savio T, Schwab ME (1990) Lesioned corticospinal tract axons regenerate in myelin-free rat spinal cord. Proc Natl Acad Sci USA 87(11):4130–4133

Saxén L, Toivonen S (1962) Primary embryonic induction. Logos, London

Schlosshauer B, Dutting D, Wild M (1990) Target-independent regulation of a novel growth associated protein in the visual system of the chicken. Development 109(2):395–409

Schnitzer J, Franke WW, Schachner M (1981) Immunohistochemical demonstration of vimentin in astrocytes and ependymal cells of developing and adult nervous system. J Cell Biol 90:435–447

Schoenfeld TA, McKerracher L, Obar R, Vallee RB (1989) MAP 1A and MAP 1B are structurally related microtubule associated proteins with distinct developmental patterns in the CNS. J Neurosci 9(5):1712–1730

Schramm LP, Adair JR, Stribling JM, Gray LP (1975) Preganglionic innervation of the adrenal gland of the rat: a study using horseradish peroxidase. Exp Neurol 49:540–553

Schreyer DJ, Jones EG (1982) Growth and target finding by axons of the corticospinal tract in prenatal and postnatal rats. Neuroscience 7:1837–1855

Schreyer DJ, Jones EHC (1983) Growing corticospinal axons by-pass lesions of neonatal rat spinal cord. Neurocience 9:31–40

Schreyer DJ, Jones EHG (1988) Topographic sequence of outgrowth of corticospinal axons in the rat: a study using retrograde axonal labeling with Fast blue. Dev Brain Res 38:89–101

Schroder HD (1980) Organization of the motoneurons innervating the pelvic muscles of the male rat. J Comp Neurol 192:567–587

Schwab ME (1990a) Myelin-associated inhibitors of neurite growth and regeneration in the CNS. TINS 13:452–456

Schwab ME (1990b) Myelin-associated inhibitors of neurite growth. Exp Neurol 109(1):2–5

Schwab ME, Schnell L (1989) Region-specific appearance of myelin constituents in the developing rat spinal cord. J Neurocytol 18:161–169

Schwab ME, Schnell L (1991) Channeling of developing rat corticospinal tract axons by myelin-associated neurite growth inhibitors. J Neurosci 11(3):709–721

Shieh JY, Leong SK, Wong WC (1983) Origin of the rubrospinal tract in neonatal, developing and mature rats. J Comp Neurol 214:79–86

Sidman RL, Rakic P (1973) Neuronal migration with special reference to developing human brain: a review. Brain Res 62:1–35

Sidman RL, Miale IL, Feder N (1959) Cell proliferation and migration in the primitive ependymal zone; an autoradiographic study of histogenesis in the nervous system. Exp Neurol 1:322–333

Sieber-Blum M (1989) SSEA-1 is a specific marker for the spinal sensory neurons lineage in the quail embryo and in neural crest cell culture. Dev Biol 134:362–375

Sillito AM, Kemp JA (1983) Cholinergic modulation of the functional organization of the cat visual cortex. Brain Res 289:143–155

Silos-Santiago I, Snider WD (1990) Development of commissural neurons in mammalian spinal cord. Abstr Neurosci 17:20.19

Silver J, Lorenz SE, Wahlsten D, Coughlin J (1982) Axonal guidance during development of the great cerebral commissures: descriptive and experimental studies in vivo on the role of preformed glial pathways. J Comp Neurol 210:10–29

Singer M (1979) Central core control of visual cortex function. In: Schmitt FO, Worden FG (eds) The neurosciences: fourth study program. MIT Press, Cambridge, pp 1093–1110

Singer M, Nordlander RH, Edgar M (1979) Axonal guidance during neurogenesis and regeneration in the spinal cord of the newt. The blueprint hypothesis of neuronal pathway patterning. J Comp Neurol 185:1–22

Skene JHP, Jacobson RD, Snipes GJ, McGuire CB, Norden JJ, Freeman JA (1986) A protein induced during nerve growth (GAP-43) is a major component of growth-cone membranes. Science 233:783–786

Smart IHM (1984) Histogenesis of the mesocortical area of the mouse telencephalon. J Anat 138:537–552

Smith CL (1983) The development and postnatal organization of primary afferent projections to the rat thoracic spinal cord. J Comp Neurol 220:29–43

Smits-van Prooije AE (1986) Processes involved in normal and abnormal fusion of the neural walls in murine embryos. Thesis, Leiden, the Netherlands

Snow DM, Steindler DA, Silver J (1990) Molecular and cellular characterization of the glial roof plate of the spinal cord and optic tectum: a possible role for a proteoglycan in the development of an axon barrier. Dev Biol 138(2):359–376

Söderholm U (1965) Histochemical localization of esterases, phosphatases and tetrazolium reductases in the motor neurons of the spinal cord of the rat and the effect of nerve division. Acta Physiol Scand 65 [Suppl 256]:1–60

Solter D, Knowles BB (1978) Monoclonal antibody defining a stage-specific mouse embryonic antigen (SSEA-1). Proc Natl Acad Sci USA 75:5565–5569

Solter D, Knowles BB (1979) Developmental stage-specific antigens during mouse embryogenesis. Curr Top Dev Biol 13:167–98

Sotelo C (1981) Development of synaptic connections in genetic and experimentally induced cerebellar malformations. In: Garrod DR, Feldman JD (eds) Development in the nervous system. University Press, Cambridge, pp 61–84

Sotelo C, Changeux J-P (1974) Transsynaptic degeneration "en cascade" in the cerebellar cortex of staggerer mutant mice. Brain Res 67:519–526

Sperry RW (1941) The effect of crossing nerves to antagonistic muscles in the hind limb of the rat. J Comp Neurol 75:1–19

Sperry RW (1963) Chemoaffinity in the orderly growth of nerve fiber patterns and connections. Proc Natl Acad Sci USA 50:703–710

Spitzer NC (1981) Development of membrane properties in vertebrates. TINS 4:169–172

Spitzer NC (1982) The development of electrical excitability. In: Sears TA (ed) Neuronal and glial interrelationships. Springer, Berlin Heidelberg New York

Sprague JM (1958) The distribution of dorsal root fibres on motor cells in the lumbosacral spinal cord of the cat, and the site of excitatory and inhibitory terminals in monosynaptic pathways. Proc R Soc Lond [Biol] 140:534–556

Stanfield BB (1989) Evidence that dorsal locus coeruleus neurons can maintain their spinal cord projection following neonatal transection of the dorsal adrenergic bundle in the rat. Exp Brain Res 78:533–538

Sterling P, Kuypers HGJM (1967) Anatomical organization of the branchial spinal cord of the cat. I. The distribution of dorsal root fibres. Brain Res 4:1–15

Swanson LW (1976) The locus coeruleus: a cytoarchitectonic, Golgi and immunohistochemical study in the albino rat. Brain Res 110:39–56

Swanson LW, Sawchenko PE, Rivier J, Vale WW (1983) Organization of ovine corticotropin releasing factor immunoreactive cells in the rat brain: an immunocytochemical study. Neuroendocrinology 36:165–186

Tascos NA, Parr J, Gonatas NK (1982) Immunocytochemical study of the glial fibrillary acidic protein in human neoplasms of the central nervous system. Hum Pathol 13:454–458

Taylor SJ, Bartlett MJ, Greenfield SA (1988) Release of acetylcholinesterase within the guinea-pig substabtia nigra: effects of 5-hydroxytryptamine and amphetamine. Neuropharmacology 27(5):507–514

Temple S (1989) Division and differentiation of isolated CNS blast cells in microculture. Nature 340:471–473

Temple S, Raff MC (1985) Differentiation of a bipotential glial progenitor cell in a single cell microculture. Nature 313:223–225

Ten Donkelaar HJ (1982) Organization of descending pathways to the spinal cord in amphibians and reptiles. Prog Brain Res 57:25–68

Tessier-Lavigne M, Placzek M, Lumsden AGS, Dodd J, Jessel TM (1988) Chemotropic guidance of developing axons in the mammalian central nervous system. Nature 336: 775–778

Tetteroo PAT, Mulder A, Lansdorp PM, Zola H, Baker DA, Visser FJ, Von dem Borne AEGK (1984) Myeloid-associated antigen 3(a)-fucosyl N-acetyl lactosamine (FAL); location on various granulocyte-membrane glycoproteins and masking upon monocytic differentiation. Eur J Immunol 14:1089–1095

Thong IG, Dreher B (1987) The development of the corticotectal pathway in the albino rat: transient projections from the visual and motor cortices. Neurosci Lett 80:275–282

Tohyama M, Satoh K, Sakumoto T, Kimoto Y, Takahashi Y, Yamamoto K, Itakura T (1978) Organization and projections of the neurons in the dorsal tegmental area of the rat. J Hirnforsch 19:165–176

Tolbert DL, Panneton PM (1983) Transient cerebrocerebellar projections in kittens: postnatal development and topography. J Comp Neurol 221:216–228

Tosney KW, Landmesser LT (1985) Specificity of motoneuron growth cone outgrowth in the chick hindlimb. J Neurosci 5:2336–2344

Tracey A (1985) Ascending and descending pathways in the spinal cord. In: Paxinos G (ed) The rat central nervous system, vol 2. Hindbrain and spinal cord. Academic, London, pp 311–324

Traub P (1985) Intermediate filaments: a review. Springer, Berlin Heidelberg New York

Tucker RP (1990) The roles of microtubule-associated proteins in brain morphogenesis: a review. Brain Res Rev 15(2):101–120

Tucker RP, Matus A (1988) Microtubule-associated proteins characteristics of embryonic brain are found in the adult mammalian retina. Dev Biol 130:423–434

Tucker RP, Binder LI, Matus A (1988) Neuronal microtubule-associated proteins in the embryonic avian spinal cord. J Comp Neurol 271:44–55

Tucker RP, Garner CC, Matus A (1989) In situ localization of microtubule-associated protein mRNA in the developing and adult rat brain. Neuron 1:1245–1256

Ueyama T, Arakawa H, Mizuno N (1987) Central distribution of efferent and afferent components of the pudendal nerve in rat. Anat Embryol (Berl) 177:37–49

Uhl G, Goodman RR, Snyder S (1979) Neurotensin-containing cell bodies, fibers and nerve terminals in the brainstem of the rat: immunohistochemical mapping. Brain Res 167:77–91

Unsicker K, Reichert-Preibsch H, Pettmann B, Labourdette G, Sensenbrenner M (1986) Astroglial and fibroblast growth factors promote in vitro survival of neurons from the peripheral and central nervous system. Soc Neurosci Abstr 12:300.7

Urdal DL, Brentnall TA, Bernstein ID, Hakamori S (1983) A granulocyte reactive mono-clonal antibody, 1G10, identifies the GA1-1-4-Fuc (a)1-3-GlcNAc (X-determinant) expressed in HL-60 cells on both glycolipid and glycoprotein molecules. Blood 62: 1022–1026

Valet J, Fulcrand J, Privat A, Marty R (1978) Radio-autographic study of cell proliferation secondary to wallerian degeneration in the postnatal rat optic nerve. Acta Neuropathol (Berl) 42:205–209

Valet J, Privat A, Fulcrand J (1983) Multiplication and differentiation of glial cells in the optic nerve of the postnatal rat. Anat Embryol (Berl) 167:335–346

Van der Loos H (1965) The "improperly" oriented pyramidal cell in the cerebral cortex and its possible bearing on problems of neuronal growth and cell orientation. Bull Johns Hopkins Hosp 117:228–250

Van der Zwet J, van Tilburg PA, Poelmann RE (1986) Oren wapperen, een betrouwbare manier om de bronst bij ratten te bepalen. Biotechnology 3:48

Van Muijen GNP, Ruiter DJ, van Leeuwen C, Prins FA, Rietsema K, Warnaar SO (1984) Cytokeratin and neurofilament in lung carcinomas. Am J Pathol 116:363–369

Van Straaten HWM, Drukker J (1987) Influence of the notochord on the morphogenesis of the neural tube. In: Wolff et al. (eds) Mesenchymal-epithelial interactions in neural development. Springer, Berlin Heidelberg New York, pp 153–162

Van Straaten HWM, Thors F, Wiertz-Hoessels L, Hekking J, Drukker J (1985) Effect of a notochordal inplant on the early morphogenesis of the neural tube and neuroblasts: histometrical and histological results. Dev Biol 110:247–254

Van Straaten HWM, Hekking JWM, Wiertz-Hoessels EJLM, Thors F, Drukker J (1988) Effect of the notochord on the differentiation of a floor plate area in the neural tube of the chick embryo. Anat Embryol (Berl) 177:317–324

Van Straaten HWM, Hekking JWM, Beursgens JPWM, Terwindt-Rouwenhorst E, Drukker J (1989) Effect of the notochord on proliferation and differentiation in the neural tube of the chick embryo. Development 107:793–803

Verburgh CA, Kuypers HGJM (1987) Branching neurons in the cervical spinal cord: a retrograde fluorescent double-labeling study in the rat. Exp Brain Res 68:565–578

Viereck C, Tucker RP, Matus A (1989) The adult rat olfactory system expresses microtubule-associated proteins found in the developing brain. J Neurosci 9:3547–3557

Von Monakow C (1883) Experimenteller Beitrag zur Kenntnis der Corpus restiforme, des "ausseren Acusticuskerns" und deren Beziehugen zum Rückenmark. Arch Psychiatr Nervenkr 14:1–16

Waibl H (1973) Zur Topographie der Medulla spinalis der Albinoratte (Rattus norvegicus). Adv Anat Embryol Cell Biol 47

Waldron HA, Gwyn DG (1969) Descending nerve tracts in the spinal cord of the rat. I. Fibres from the midbrain. J Comp Neurol 137:143–154

Walicke P, Cowan WM, Ucno N, Baird A, Guillemin R (1986) Fibroblast growth factor promotes survival of dissociated hippocampal neurons and enhances neurite extension. Proc Natl Acad Sci USA 83:3012–3016

Walker AE, Weaver TA (1942) The topical organization and termination of the fibers of the posterior columns in Macaca mulatta. J Comp Neurol 76:145–158

Wallace JA, Lauder JM (1983) Development of the serotoninergic system in the rat embryo: an immunohistochemical study. Brain Res Bull 10:459–479

Walter J, Muller B, Bonhoeffer F (1990) Axonal guidance by an avoidance mechanism. J Physiol (Paris) 84(1):104–110

Warf BC, Fok-Seang J, Miller RH (1991) Evidence for the ventral origin of oligodendrocyte precursors in the rat spinal cord. J Neurosci 11:2477–2488

Webb CP, Greenfield SA (1992) Non-cholinergic effects of acetylcholinesterase in the substantia nigra: a possible role for an ATP-sensitive potassium channel. Exp Brain Res 89:49–58

Weiss P (1924) Die Funktion transplantierter Amphibien-Extremitäten. Aufstellung einer Resonanztheorie der motorischen Nerventätigkeit auf Grund abgestimmter Endorgane. Arch Mikrosk Anat Entwicklungsmech 102:635–672

Weiss P (1934) In vitro experiments on the factors determining the course of the outgrowing nerve fiber. J Exp Zool 68:393–448

Wessels WJT (1991) Development of the peripheral sensory innervation of the rat hindlimb and its central connections: a WGA-HRP study. Thesis, Leiden, the Netherlands

Wessels WJT, Feierabend HKP, Marani E (1990) Evidence for a rostrocaudal organization in dorsal root ganglia development as demonstrated by intra-uterine WGA-HRP injections into the hindlimb of rat fetuses. Dev Brain Res 54:273–281

Westlund KN, Bowker RM, Ziegler MG, Coulter JD (1981) Origins of spinal noradrenergic pathways demonstrated by retrograde transport of antibody to dopamine-1-hydroxylase. Neurosci Lett 25:243–249

Westlund KN, Bowker RM, Ziegler MG, Coulter JD (1983) Noradrenergic projections to the spinal cord of the rat. Brain Res 263:15–31

Wictorin K, Brundin P, Gustavii B, Lindvall O, Bjorklund A (1990) Reformation of long axon pathways in adult rat central nervous system by human forebrain neuroblasts. Nature 347:556–558

Williams BP, Abney ER, Raff MC (1985) Macroglial cell development in embryonic rat brain: studies using monoclonal antibodies, fluorescence activated cell sorting and cell culture. Dev Biol 112:126–134

Windle WF, Austin MF (1936) Neurofibrillar development in the central nervous system of chick embryos up to five days incubation. J Comp Neurol 63:431–463

Windle WF, Baxter RE (1936) The first neurofibrillar development in albino rat embryos. J Comp Neurol 63:173–187

Xu XM, Martin GF (1989) Developmental plasticity of the rubrospinal tract in the North American opossum. J Comp Neurol 279:368–381

Yamada KM, Spooner BS, Wessels NK (1970) Axon growth: roles of microfilaments and microtubules. Proc Natl Acad Sci USA 66:1206–1212

Yamamoto M, Boyer AM, Schwarting A (1985) Fucose-containing glycolipids are stage- and region-specific antigens in developing embryonic brain of rodents. Proc Natl Acad Sci USA 82:3045–3049

Zemlan FP, Leonard CM, Kow L-M, Pfaff DW (1978) Ascending tracts of the lateral columns of the rat spinal cord: a study using the silver impregnation and horseradish peroxidase techniques. Exp Neurol 62:298–334

Springer-Verlag and the Environment

We at Springer-Verlag firmly believe that an international science publisher has a special obligation to the environment, and our corporate policies consistently reflect this conviction.

We also expect our business partners – paper mills, printers, packaging manufacturers, etc. – to commit themselves to using environmentally friendly materials and production processes.

The paper in this book is made from low- or no-chlorine pulp and is acid free, in conformance with international standards for paper permanency.